NanoScience and Technology

The series NanoScience and Technology is focused on the fascinating nano-world, mesoscopic physics, analysis with atomic resolution, nano and quantum-effect devices, nanomechanics and atomic-scale processes. All the basic aspects and technology-oriented developments in this emerging discipline are covered by comprehensive and timely books. The series constitutes a survey of the relevant special topics, which are presented by leading experts in the field. These books will appeal to researchers, engineers, and advanced students.

More information about this series at http://www.springer.com/series/3705

Alev Devrim Güçlü · Pawel Potasz
Marek Korkusinski · Pawel Hawrylak

Graphene Quantum Dots

Springer

Alev Devrim Güçlü
Department of Physics
Izmir Institute of Technology
Izmir
Turkey

Pawel Potasz
Institute of Physics
Wrocław University of Technology
Wrocław
Poland

Marek Korkusinski
Emerging Technologies Division, Quantum Theory Group
National Research Council of Canada
Ottawa, ON
Canada

Pawel Hawrylak
Department of Physics
University of Ottawa
Ottawa, ON
Canada

ISSN 1434-4904 ISSN 2197-7127 (electronic)
ISBN 978-3-662-44610-2 ISBN 978-3-662-44611-9 (eBook)
DOI 10.1007/978-3-662-44611-9

Library of Congress Control Number: 2014947690

Springer Heidelberg New York Dordrecht London

Printed on acid-free paper

Springer is part of Springer Science+Business Media (www.springer.com)

Preface

When one of us, PH, arrived at the University of Kentucky to start his Ph.D. with K. Subbaswamy in 1981, graphene in intercalated graphite (GIC) was all the rage. He was given a paper by Wallace describing electronic properties of graphene and graphite and told to go and talk to Peter Eklund's group who was measuring optical properties of intercalated graphite next door. The next 4 years were exciting, with the standing room only at the graphite sessions at the March Meetings, it seemed that future belonged to graphene. However, the excitement did not last forever, and after completing Ph.D. PH went on to work on another class of artificially made materials, semiconductor heterostructures. The last 30 years has seen the ability of controlling semiconductors moving from heterojunctions and superlattices to three-dimensional control and making semiconductor quantum dots. Today, semiconductor quantum dots enable, for example, transistors based on spins of single electrons, sources of single and entangled photons, efficient quantum dot lasers, biomarkers, and solar cells with improved efficiency.

In this monograph, we describe a new class of quantum dots based on graphene, a single atomic layer of carbon atoms. Since the isolation of a single graphene layer by Novoselov and Geim, we became interested in using only graphene, instead of different semiconductors, to create graphene quantum dots. By controlling the lateral size, shape, type of edge, doping level, sublattice symmetry, and the number of layers we hoped to engineer electronic, optical, and magnetic properties of graphene. Our initial exploration started in 2006, but came into focus later after we became aware of a beautiful work by Ezawa and by Palacios and Fernandez-Rossier on triangular graphene quantum dots. This work emphasized the role of sublattice symmetries and electron-electron interactions in engineering magnetic properties of graphene nanostructures, opening the possibility of creating an interesting alternative to semiconductor spintronics. The second intriguing possibility offered by graphene is that it is a semimetal with zero-energy gap. By lateral size quantization the gap in graphene quantum dots can be tuned from zero to UV. By contrast, in semiconductors, the energy gap can only be larger than the energy gap of the bulk material. In principle, graphene quantum dots allow for design of material with the desired energy gap. The exciting possibility of convergence and seamless integration of electronics, photonics, and spintronics in a single material, graphene, could lead to a new area of research, carbononics.

These were some of the ideas we embarked to explore when two of us, ADG and PP joined the Quantum Theory Group led by PH at the NRC Institute for Microstructural Sciences in 2008. The monograph is based largely on the Ph.D. thesis of one of us, Pawel Potasz, shared between NRC and Wrocław University of Technology.

After Introduction in Chap. 1, Chap. 2 describes the electronic properties of bulk graphene, a two dimensional crystal, including fabrication, electronic structure, and effects of more than one layer. In Chap. 3 fabrication of graphene quantum dots is described while Chap. 4 describes single particle properties of graphene quantum dots, including tight-binding model, effective mass, magnetic field, spin-orbit coupling, and spin Hall effect. The role of sublattice symmetry and the emergence of a degenerate shell of electronic states in triangular graphene quantum dots is described. The bilayers and rings, including Möbius ring with topology encoded by geometry, are described. Chapter 5 introduces electron-electron interactions, including introduction to several tools such as Hartree–Fock, Hubbard model and Configuration Interaction method used throughout the monograph. Chapter 6 discusses correlations and magnetic properties in triangular graphene quantum dots and rings with degenerate electronic shells, including existence of magnetic moment and its melting with charging, and Coulomb and Spin Blockade in transport. Chapter 7 focuses on optical properties of graphene quantum dots, starting with tight-binding model and including self-energy and excitonic corrections. Optical spin blockade and optical control of the magnetic moment is described. Comparison with experimental results obtained for colloidal graphene quantum dots is also included.

We hope the monograph will introduce the reader to this exciting and rapidly evolving field of graphene quantum dots and carbononics.

Izmir, Turkey
Wrocław, Poland
Ottawa, Canada

Alev Devrim Güçlü
Pawel Potasz
Marek Korkusinski
Pawel Hawrylak

Contents

Chapter 1
Introduction

Abstract This chapter introduces and motivates the subject of the monograph, the rapidly growing field of research on the electronic, optical and magnetic properties of graphene quantum dots.

Graphene is a one-atom thick two-dimensional crystal of carbon atoms. Weakly bound planes of graphene form graphite with electronic properties engineered by intercalation [1], and rolled and folded graphene is a building block of fullerenes and carbon nanotubes [2].

Since the isolation of a single layer of graphene [3–6] and the demonstration of its excellent conductivity and optical properties, the research aiming at determining the electronic properties and potential applications of graphene progressed at a rapid pace. Much of the current understanding of the electronic properties of graphene has been reviewed by Castro-Neto et al. [7], transport properties by Das Sarma et al. [8] and many-body effects by Kotov et al. [9], Vozmedano et al. [10] and MacDonald et al. [11]. An excellent overview of many aspects of graphene, from chemistry to fundamental problems in quantum matter, can be found in a series of articles in the Proceedings of the Nobel Symposium 148 [12] on "Graphene and quantum matter" celebrating the 2010 Noble Prize in Physics for graphene for Geim and Novoselov. An extensive introduction to graphene can also be found in books by, e.g., Katsnelson [13], Aoki et al. [14] and Torres et al. [15].

The list of some of the exciting properties of graphene starts with graphene being an ideal, only one atom thick, two-dimensional crystal. Because graphene is built of carbon, pure graphene is free of nuclear spins and should be an attractive material for electron-spin based quantum circuits. However, carbon atom has no magnetic moment, hence realizing magnetism in graphene is challenging. The linear dispersion of quasiparticles in graphene, Dirac Fermions, leads to a number of interesting effects. The two-sublattice structure of graphene couples Dirac Fermions with sublattice index, pseudospin, and introduces Berry's phase. The relativistic-like effects lead to Klein tunneling and absence of electrostatic confinement. The interaction among Dirac Fermions is different from the interaction among Schrödinger electrons and plays an important role in determining the electronic properties of graphene. The role of interactions in, e.g., renormalization of Fermi velocity continues to be a subject of intense research.

A.D. Güçlü et al., *Graphene Quantum Dots*,
NanoScience and Technology, DOI 10.1007/978-3-662-44611-9_1

Given these interesting electronic properties and much progress in our understanding of graphene, a new challenge emerges: Can we take graphene as a starting material and engineer its electronic, optical and magnetic properties by controlling the lateral size, shape, type of edge, doping level, and the number of layers in "graphene quantum dots"? Graphene is a semimetal, i.e., it has no gap. By controlling the lateral size of graphene the energy gap can be tuned from THz to UV covering entire solar spectrum, the wavelength needed for fiber based telecommunication (telecom window) and THz spectral range. One can also envision building a magnet, a laser, and a transistor using carbon material only and creating disposable and flexible nanoscale quantum circuits out of graphene quantum dots [16]. The research on graphene quantum dots is rapidly expanding covering physics, chemistry, materials science, and chemical engineering. This monograph attempts to present the current understanding of graphene quantum dots. An attempt is made to cover the rapidly expanding and evolving field but the monograph focuses mainly on the work done at the Institute for Microstructural Sciences, National Research Council of Canada. The authors thank I. Ozfidan, O. Voznyy, E. Kadantsev, C.Y. Hsieh, A. Sharma and A. Wojs for their contributions.

References

1. M.S. Dresselhaus, G. Dresselhaus, Intercalation compounds of graphite. Advances in Physics **30**(2), 139–326 (1981)
2. M.S. Dresselhaus, Phys. Scr. **T146**, 014002 (2012)
3. K.S. Novoselov, A.K. Geim, S.V. Morozov, D. Jiang, Y. Zhang, S.V. Dubonos, I.V. Grigorieva, A.A. Firsov, Science **306**, 666 (2004)
4. K.S. Novoselov, A.K. Geim, S.V. Morozov, D. Jiang, M.I. Katsnelson, I.V. Grigorieva, S.V. Dubonos, A.A. Firsov, Nature **438**, 197 (2005)
5. Y. Zhang, Y.W. Tan, H.L. Stormer, P. Kim, Nature **438**, 201 (2005)
6. M.L. Sadowski, G. Martinez, M. Potemski, C. Berger, W.A. de Heer, Phys. Rev. Lett. **97**, 266405 (2006)
7. A.H. Castro Neto, F. Guinea, N.M.R. Peres, K.S. Novoselov, A.K. Geim, Rev. Mod. Phys. **81**, 109 (2009)
8. S. Das Sarma, S. Adam, E.H. Hwang, E. Rossi, Rev. Mod. Phys. **83**, 407 (2011)
9. V.N. Kotov, B. Uchoa, V.M. Pereira, F. Guinea, A.H. Castro Neto, Rev. Mod. Phys. **84**, 1067–1125 (2012)
10. M.A.H. Vozmediano, F. Guinea, Phys. Scr. **T146**, 014015 (2012)
11. A.H. MacDonald, J. Jung, F. Zhang, Phys. Scr. **T146**, 014012 (2012)
12. A. Niemi, F. Wilczek, E. Ardonne, H. Hansson, Phys. Scr. **T146**, 010101 (2012)
13. M.I. Katsnelson, *Graphene: Carbon in Two Dimensions* (Cambridge University Press, Cambridge, 2012)
14. H. Aoki, M.S. Dresselhaus (eds.), *Physics of Graphene* (Springer, Heidelberg, 2014)
15. L.E.F. Foa Torres, S. Roche, J.-C. Charlier, *Introduction to Graphene Based Nanomaterials: From Electronic Structure to Quantum Transport* (Cambridge University Press, Cambridge, 2014)
16. A.D. Güçlü, P. Potasz, P. Hawrylak, Graphene-based integrated electronic, photonic and spintronic circuit, invited paper, in *Future Trends in Microelectronics 2012*, ed. by S. Luryi, J. Xu, A. Zaslavsky (Wiley, New York, 2013), p. 308

Chapter 2
Graphene—Two-Dimensional Crystal

Abstract After a brief review of the history of research on carbon materials, this chapter describes fabrication methods, mechanical properties and electronic band structure of bulk graphene, including the tight-binding model, effective mass model of Dirac Fermions, Berry's phase, chirality and absence of backscattering, and the effect of interlayer coupling on bilayer graphene.

2.1 Introduction to Graphene

Graphene is a one-atom thick planar structure of carbon atoms arranged in a honeycomb crystal lattice. It is a basis for the understanding of the electronic properties of other allotropes of carbon. Graphene can be stacked up to form a 3D crystal of graphite, rolled up along a given direction to form nanotubes [1], an example of 1D material, or wrapped up into a ball creating fullerene, an example of 0D material [2]. It is worth to note that the 1996 Nobel Prize in Chemistry was awarded jointly to Robert F. Curl Jr., Sir Harold W. Kroto and Richard E. Smalley "for their discovery of fullerenes", the 2010 Nobel Prize in Physics was awarded to Andre Geim and Konstantin Novoselov for their "groundbreaking experiments regarding the two-dimensional material graphene", and the 2012 Kavli Prize in Nanoscience to Mildred Dresselhaus "for her pioneering contributions to the study of phonons, electron-phonon interactions, and thermal transport in nanostructures", mainly carbon based materials.

Research on graphene has a long history. One of the first papers was written by P.R. Wallace in 1946 at the National Research Council of Canada [3] Chalk River Laboratory. It described a band structure of graphite, starting with a single layer—graphene. Wallace correctly identified the structure of graphene layer with two non-equivalent carbon sublattices, and described and solved a tight-binding model of graphene. Wallace demonstrated that the conduction and valence bands of graphene touch at two non-equivalent points of the Brillouin zone and hence that graphene is a semimetal with an unusual linear dispersion of quasi-particle energy as a function of the wave vector. This behavior is in close analogy to the dispersion of massless relativistic particles as described by the Dirac and Weyl equations [4, 5]

A.D. Güçlü et al., *Graphene Quantum Dots*,
NanoScience and Technology, DOI 10.1007/978-3-662-44611-9_2

and electrons in graphene are called Dirac electrons. It took almost 60 years to directly detect Dirac Fermions in graphene [6]. A theory of the electronic properties of graphite was further developed by, e.g., Slonczewski, McClure and Weiss [7, 8] and by Dresselhaus [9]. The analogy between graphene and relativistic effects was further explored by Semenoff [10] and Haldane [11] who discussed an analogy of graphene to $(2 + 1)$ dimensional quantum electrodynamics (QED).

In the 70s and 80s much effort went into modifying the electronic properties, in particular improving conductivity of graphite by intercalation with, e.g., alkali metals resulting in graphite intercalation compounds (GIC) [12]. With intercalant atoms and molecules, e.g., Li or H_2SO_4, in-between graphene layers, the graphene layers were both effectively separated from each other and their carrier concentration was changed by either additional electrons or holes [12–15]. Hence intercalation in graphite is equivalent to doping in semiconductors, with carriers donated to graphene layers scattered by ionized impurities. The main difference between bulk semiconductors and graphite at low dopant (intercalant) concentration is the formation of stages, for example in stage two GIC intercalant is found between every second graphene layer. The intercalant in stages two and higher forms lateral domains inhibiting transport in the plane [12, 16, 17]. The electronic properties of graphite intercalation compounds were studied by a number of groups [12, 18–20]. Theory of optical properties of graphene was developed by Blinowski et al. [21] and the theory was compared with experiment [14, 21]. Effects of electron-electron interactions and collective excitations, plasmons, were also studied [22–24].

In the 80s and 90s new forms of carbon were discovered, fullerenes by Kroto et al. [2] and carbon nanotubes by Ijima et al. [1]. These major developments stimulated research on nanostructured graphene.

Graphite monolayers, graphene, were observed already in 1962 by Boehm et al. [25]. Boehm obtained thin graphite fragments of reduced graphite oxide identifying some of them as graphene (the name graphene for monolayer was introduced later, in 1986 [26]). Ultrathin graphitic films were also obtained using different growth techniques [27–30]. Analysis of their electronic properties was carried out by surface science techniques. Carrier dynamics in few-nm-thick graphite films was studied in the 90s [31, 32]. Ohashi reported resistivity changes by up to 8 % with varying electric field for 20 nm thick samples. Using bottom-up techniques, a group lead by Mullen created "giant hydrocarbons" [33, 34].

In 1999, Ruoff et al. developed a method called "mechanical exfoliation" [35]. They used a tip of the atomic force microscope (AFM) to manipulate small pillars patterned in the highly oriented pyrolytic graphite (HOPG) by plasma etching, Fig. 2.1. HOPG is characterized by high atomic purity and smooth surface. Carbon layers could be delaminated due to the weak van der Waals forces between consecutive layers. The mechanical exfoliation method was realized by Geim's group using scotch tape. In 2004 Geim and co-workers exfoliated a few carbon layers from graphite, deposited them on silicon transistor structure and showed ambipolar electric field effect in thin graphene flakes at ambient conditions [36] (Fig. 2.2). In parallel, de Heer and co-workers obtained few-layer graphene on the surface of silicon carbide [37]. The method of identifying only a few layers in graphene samples fabricated

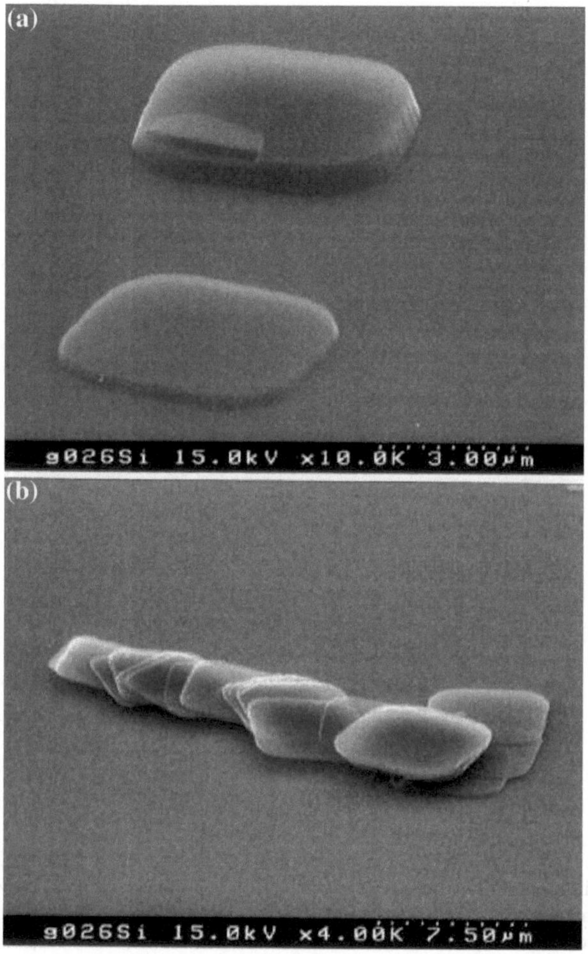

Fig. 2.1 SEM images of thin graphite plates on the Si(001) substrate. Reprinted from [35]

using scotch-tape technique required a combination of optical microscope (OM), scanning electron microscope (SEM) and AFM. Thin graphite fragments, thinner than 50 nm, were completely invisible in OM but clearly seen in high-resolution SEM on SiO_2 substrate, Fig. 2.3. The optical path added by graphene layers shifted the interference colors from violet-blue for pure SiO_2 substrate to blue for samples with graphitic films. These color shifts turned ou to be sensitive to the number of graphene layers. A contrast was affected by the thickness of the SiO_2 substrate and the best contrast was obtained for 300 nm thick substrate. The thickness of the substrate was crucial because 5 % change in substrate thickness can make graphene completely invisible. After a first selection of thinnest fragments, AFM was used to identify fragments with thickness less than ~1.5 nm because they were invisible

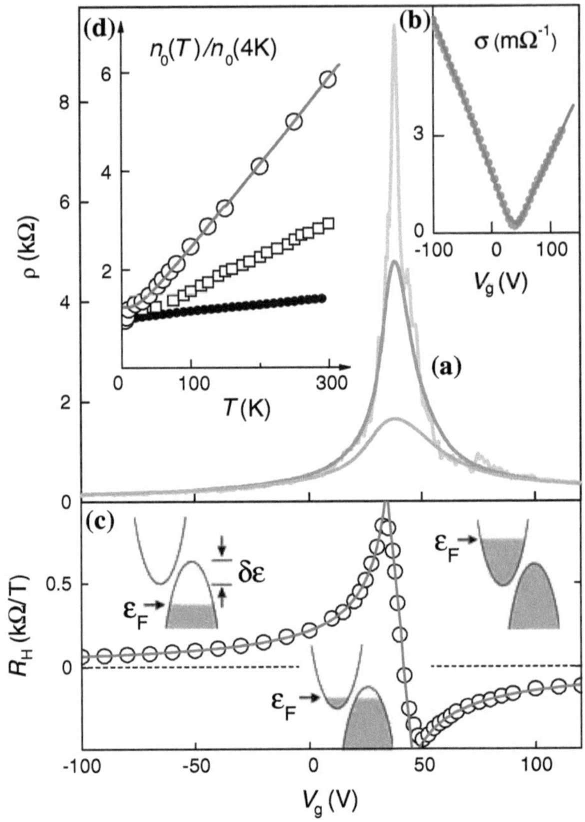

Fig. 2.2 Electric field effect in thin graphene flakes. **a** Typical dependences of FLGs resistivity ρ on gate voltage for different temperatures (T = 5, 70, and 300 K for *top* to *bottom curves*, respectively). **b** Example of changes in the film's conductivity $\sigma = 1/\rho(V_g)$ obtained by inverting the 70 K curve (*dots*). **c** Hall coefficient R_H versus Vg for the same film; T = 5 K. **d** Temperature dependence of carrier concentration n_0 in the mixed state for the film in (**a**) (*open circles*), a thicker FLG film (*squares*), and multilayer graphene ($d \simeq 5$ nm; *solid circles*). *Red curves* in **b–d** are the dependences calculated from proposed model of a 2D semimetal illustrated by insets in (**c**). Reprinted from [36]

even via the interference shift, Fig. 2.4. Later, a group lead by Geim has shown a simple method of distinguishing single layer graphene, even with respect to bilayer, by using Raman spectroscopy [38]. The exfoliated samples were characterized by high carrier mobility, exceeding 10,000 cm^2/Vs, at ambient conditions. The high mobility was crucial for the observation of ballistic transport over submicron distances. It was shown that in thin graphene flakes a perpendicular electric field changed resistivity by a factor of ~100. The change in resistivity was attributed to variable carrier density as in silicon-based field-effect transistors, an effect which cannot be realized in metallic conductors. It was also shown that independently of carrier concentration, the graphene conductivity was larger than a minimum value corresponding

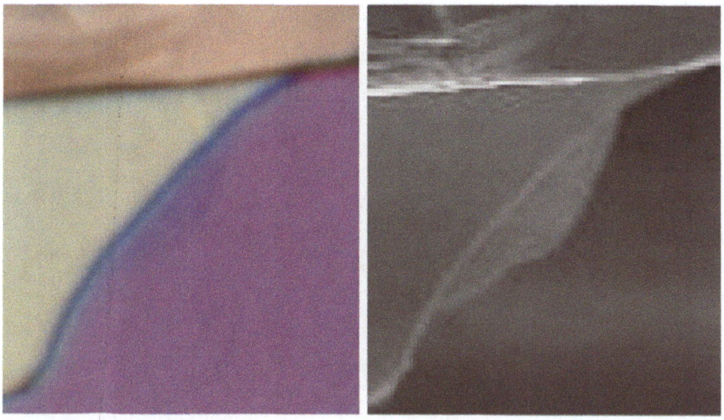

Fig. 2.3 Images of a thin graphitic flake in optical (*left*) and scanning electron (*right*) microscopes. Few-layer graphene is clearly visible in SEM (in the *center*) but not in optics. Reprinted from supporting materials of [36]

to the quantum unit of conductance [36, 39]. Perhaps the most surprising in their experiment [36] was not the observation and the isolation of graphene but measured high conductivity [40]. This implied that atomic planes remained continuous and conductive even when exposed to air, i.e., under ambient conditions.

The first experiments were followed by experiments on a single graphene layer by Geim's and Kim's groups [39, 41]. Based on magneto-transport measurements, a single layer was shown to indeed exhibit a linear energy dispersion, confirmed later by photoemission experiments [6].Integer quantum Hall effect (IQHE) in graphene is different from that in conventional semiconductors with a parabolic dispersion as will be discussed later on. In graphene, Hall plateaus appear at half-integer filling factors with Landau level dispersion proportional to the square root of the magnetic field, Fig. 2.5.

Additionally, the unit of quantized conductance is 4 times larger than in conventional semiconductors. This is related to fourfold degeneracy in graphene (spin degeneracy and valley degeneracy). In 2007, IQHE in graphene was demonstrated at room temperature [42, 43]. This was possible due to a high quality of samples and large cyclotron energies of "relativistic" electrons, and consequently a large separation between neighboring lowest Landau levels, Fig. 2.6.

The relativistic nature of carriers in graphene is also interesting from fundamental point of view. Electrons close to the Fermi level move like photons, with no rest mass and velocity 300 times smaller than the speed of light [44]. Thus, one can probe quantum electrodynamics (QED) in the solid state. One of the effects characteristic for relativistic particles is Klein tunneling [45, 46], Fig. 2.7. A relativistic particle can travel through a high potential barrier, in some cases with 100 % probability. This is related to the fact that a barrier for electrons is a well for holes, resulting in hole bound states inside it. Matching between electron and hole wavefunctions increases the probability of tunneling through the barrier [45]. Klein tunneling has important

Fig. 2.4 Single-layer graphene visualized by AFM. Narrow ($\simeq$100 nm) graphene stripe next to a *thicker area*. Colors: *dark brown* corresponds to SiO_2 surface, *bright orange* $\sim$2 nm, *light brown* $\sim$0.5 nm—the high of a single layer. Reprinted from supporting materials of [36]

consequences; carriers cannot be spatially confined by an electric field produced by a metallic gate. Klein tunneling in graphene was confirmed experimentally in 2009 [47, 48].

The relativistic nature of quasiparticles in graphene plays an important role in many-body effects in graphene, reviewed extensively, e.g., by Kotov et al. [49]. Unlike in a 2D gas of Schrödinger electrons, Dirac electrons have both the kinetic energy $\sim 1/\lambda$ and Coulomb energy $\sim 1/\lambda$, where λ is a characteristic length related to average interparticle separation, and the ratio of kinetic to interaction energy does not depend on carrier density but rather on external screening. Hence the effects of electron-electron interactions can be controlled not by carrier density but by external environment. From the microscopic lattice point of view, extensive Monte-Carlo calculations for a Hubbard model on a honeycomb lattice [50, 51] point to a stable semi-metallic phase for weak interactions and Mott-insulating phase at higher interactions.

Graphene interacts with light. The study of optical properties of graphene started with investigation of optical properties of graphite intercalation compounds by

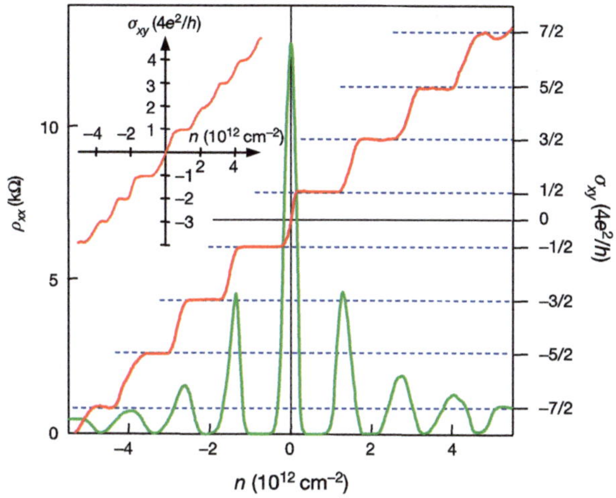

Fig. 2.5 Hall conductivity σ_{xy} (*red line*) and longitudinal resistivity ρ_{xx} (*green line*) of graphene as a function of their concentration at $B = 14$ T and $T = 4$ K. $\sigma_{xy} = (4e^2/h)\nu$ is calculated from the measured dependences of $\rho_{xy}(V_g)$ and $\rho_{xy}(V_g)$ as $\sigma_{xy} = \rho_{xy}/(\rho_{xy}^2 + \rho_{xx}^2)$. The behavior of $1/\rho_{xy}$ is similar but exhibits a discontinuity at $V_g \simeq 0$, which is avoided by plotting σ_{xy}. Inset: σ_{xy} in two-layer graphene where the quantization sequence is normal and occurs at integer ν. The latter shows that the half-integer QHE is exclusive to ideal graphene. Reprinted from [39]

Blinowski et al. [21] and Eklund et al. [14]. In n- or p-type doped GIC the filling of Dirac Fermion band resulted in blocking of absorption for photons with energy less than twice the Fermi energy. The isolation of a single layer and control over the carrier density and the Fermi level allowed for gate controlled optical properties [52, 53] and for direct observation of Dirac Fermions using photoemission spectroscopy [6]. Moreover, it was possible to measure the absorption spectrum of graphene and determine that in the photon energy range where electronic dispersion is linear, graphene suspended in air absorbs 2.3 % of incident light [54]. This implies that the absorption coefficient for single-layer graphene is several orders of magnitude higher than similar layers of semiconductors such as GaAs or germanium at 1.5 μm [55]. In parallel to experiments, progress in theory of optical properties using many-body perturbation theory GW+BSE has been reported by Louie and co-workers [56]. The possibility of controlling resistivity in a wide range, high mobility, good crystalline quality and planar structure compatible with top-down processing makes graphene an interesting material for electronic applications [57–61]. Recent experiments on suspended graphene have shown mobility as large as 200,000 cm^2/Vs which is more than 100 times larger than that of silicon transistors [62–65]. The mobility remains high even in high electric fields. The mean-free path in a suspended sample after annealing achieves 1 μm, which is comparable with a sample size. Furthermore, suspended graphene absorbs only 2.3 % of incident white light making it a useful material for transparent electrodes for touch screens and light panels [54]. Thus, graphene can be a competitor to the industrial transparent electrode material, indium

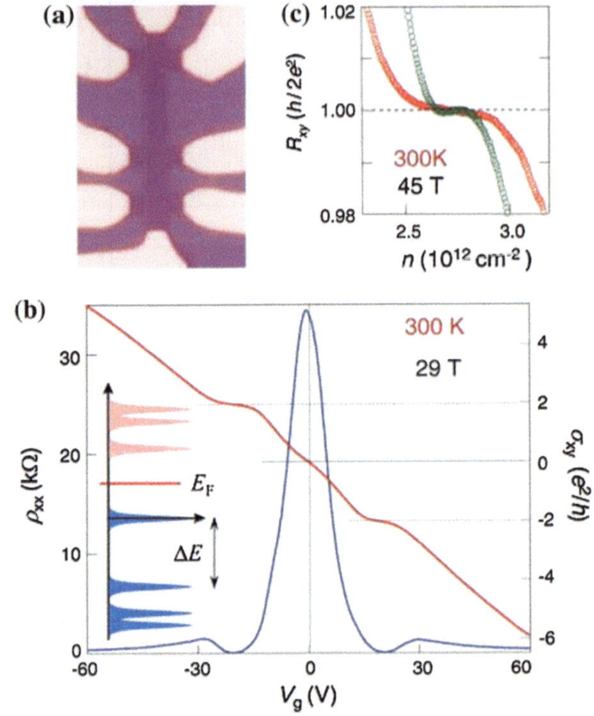

Fig. 2.6 Room-temperature QHE in graphene. **a** Optical micrograph of one of the devices used in the measurements. The scale is given by the Hall bars width of 2 μm. B σ_{xy} (*red*) and ρ_{xx} (*blue*) as a function of gate voltages (V_g) in a magnetic field of 29 T. Positive values of V_g induce electrons, and negative values of V_g induce holes, in concentrations $n = (7.2 \times 10^{10}\,\text{cm}^{-2}\text{V}^{1})V_g$ (5, 6). (Inset) The LL quantization for Dirac fermions. **c** Hall resistance, R_{xy}, for electrons (*red*) and holes (*green*) shows the accuracy of the observed quantization at 45 T. Reprinted from [42]

tin oxide (ITO) [66]. The reader may consult, e.g., an article by Avouris et al. for more information on graphene applications in electronics and photonics [55].

Some potential applications in quantum information processing were also proposed. Graphene is built of carbon atoms. ^{12}C atom does not have a finite nuclear spin and, as in light atoms, graphene has a very weak spin-orbit coupling. Hence it is expected that the electron spin will have a very long coherence time. Thus, it is a viable material for spin qubits [67, 68].

For more immediate applications, graphene can be used for gas sensors. Graphene has a maximum ratio of the surface area to volume. In typical 3D materials, resistivity is not influenced by adsorption of a single molecules on their surface. This is not true in graphene. Adsorption of molecules from surrounding atmosphere causes doping of graphene by electrons or holes depending on the nature of the gas. This can be detected in resistivity measurements [69]. Another potential application of graphene might be as a subnanometer trans-electrode membrane for sequencing DNA [70].

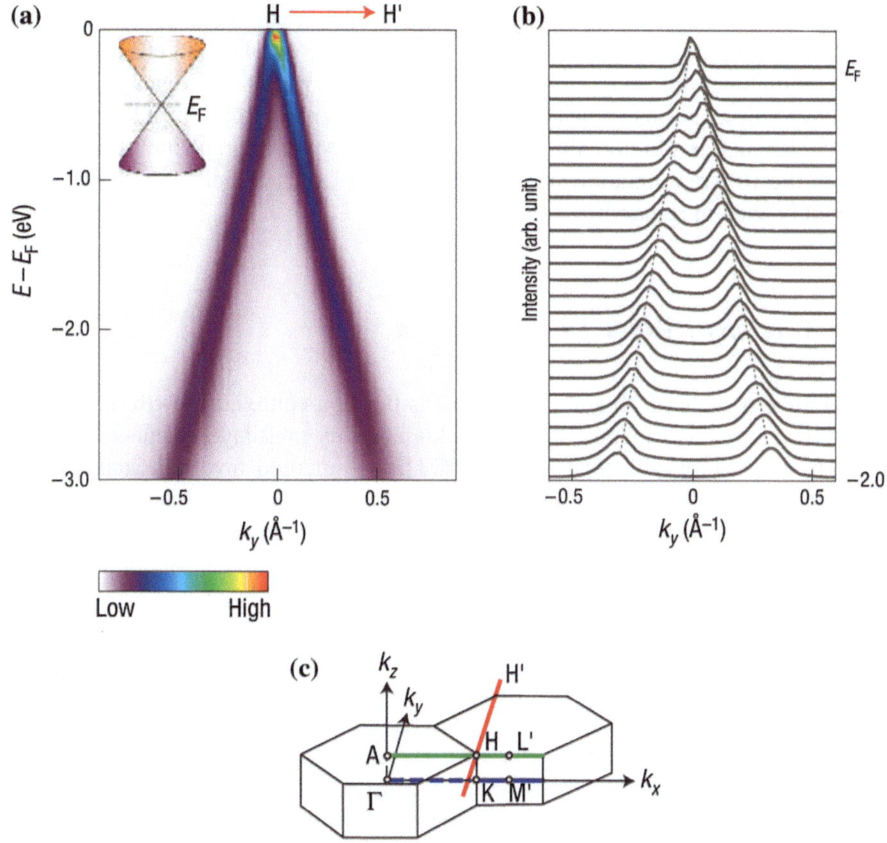

Fig. 2.7 Direct observation of linear energy dispersion near the Fermi level of graphene using photoemission spectroscopy ARPES. Reprinted from [6]

2.2 Fabrication of Graphene

Below, we describe several methods for fabrication of graphene devices and large scale growth of graphene layers.

2.2.1 *Mechanical Exfoliation*

The method used by Geim and co-workers to obtain graphene is called mechanical exfoliation [36].Graphite consists of parallel graphene sheets, weakly bound by van der Waals forces. These forces can be overcome with an adhesive tape. Novoselov, Geim and co-workers successively removed layers from a graphite flake by repeated peeling [36]. Next, graphite fragments were pressed down against a substrate leaving thin films containing down to a single layer. Due to an interference effect related to a special thickness of SiO_2 substrate (300 nm), it was possible to distinguish a few, down to a single, graphene layers, indicated by darker and lighter shades of purple. The mechanical exfoliation allows isolation of high-quality graphene samples with sizes in the 10 μm range, too small for applications such as field effect transistors, but widely used in research.

2.2.2 Chemical Vapor Decomposition

The controlled way of obtaining graphene is through epitaxial growth of graphitic layers on a surface of metals. It provides high-quality multilayer graphene samples strongly interacting with their substrate [71]. One method involves catalytic metals such as nickel, ruthenium, platinum and iron. These metals disassociate carbon precursors, e.g., CH_4, as well as dissolve significant amounts of carbon at high temperature. Upon cooling, the carbon segregates on a metal surface as graphene layer. For example, a method of growing few layer graphene films by using chemical vapor deposition (CVD) on thin nickel layers was demonstrated [58, 72]. It was shown that the number of graphene layers can be controlled by changing the nickel thickness or growth time. Transport measurements in high magnetic fields showed the half-integer quantum Hall effect, characteristic for monolayer graphene [58]. Their samples revealed good optical, electrical and mechanical properties. The sample size exceeded $1 \times 1\,\mathrm{cm}^2$ with graphene domain sizes between 1 and 20 μm. Size of graphene films was limited by CVD chamber size. It was possible to transfer the graphene layer to an arbitrary substrate, e.g., by using dry-transfer process.

The second and popular method involves catalytic CVD process where the precursor is decomposed at elevated temperature on copper foil [73, 74] and graphene is formed upon cooling. This technique yields primarily a single graphene layer approaching wafer scale crystal quality [74]. Upon dissolution of copper, graphene can be transferred to other substrates.

2.2.3 Thermal Decomposition of SiC

When SiC wafers are heated, the Si desorbs and the remaining carbon rebonds to form one or more layers of graphene on top of SiC. By using this technique, Berger, de Heer and co-workers produced few layers of graphene [37, 75]. Their samples were continuous over several mm revealing presence of the 2D electron gas with high mobility. One of the advantages of this method is the possibility of patterning films into narrow ribbons or other shapes by using conventional lithographic techniques [76–78, 80]. Additionally, insulating SiC substrates can be used, so a transfer to another insulator is not required. Emtsev et al. have improved this technique by using argon gas under high pressure [79]. The graphitization in the argon atmosphere enabled increase of processing temperature resulting in producing much larger domains of monolayer graphene and reducing the number of defects. Emtsev et al. obtained arrays of parallel terraces up to 3 μm wide and more than 50 μm long. They reported carrier mobility values only 5 times smaller than that for exfoliated graphene on substrates in the limit of high doping.

Graphene was also epitaxially grown by CVD on SiC [81–83]. The advantage of this method is that CVD growth is less sensitive to SiC surface defects. The high quality of graphene was confirmed by several techniques [83]. Single atomic layer could be identified by ellipsometry with high spatial resolution. The annealing time and argon pressure are responsible for the growth kinetics of graphene and influence the number of graphene layers. The properties of this material were studied by STM and TEM [81]. The first carbon layer was about 2 Å from the SiC surface as a result of strong covalent bonds between carbon layer and silicon atoms on the SiC surface. Creation of edge dislocations in the graphene layers as a result of bending of graphene planes on atomic steps was observed [81]. The conductivity of graphene thin films on SiC substrates was also measured [82].

2.2.4 Reduction of Graphite Oxide (GO)

In this method, graphite is chemically modified to produce graphite oxide (GO) by using the Hummer's method [84]. GO is dispersed in a solvent, e.g., water, and can be chemically exfoliated. Graphene sheets are obtained by a chemical, thermal or electrochemical reduction process of oxygen groups [85–88]. The level of oxidization determines electrical conductivity and optical transparency [89]. During this process, the quality of samples is significantly reduced due to a change from sp^2 to sp^3 hybridization for many carbon atoms resulting in decreasing mobility. On the other hand, films reveal high flexibility and stiffness much better than that of other paper-like materials [86]. The production technique is low-cost and can be scaled up to produce large pieces of graphene.

2.3 Mechanical Properties

Graphene is a two-dimensional crystal continuous on a macroscopic scale [90]. Surprisingly, it is stable under ambient conditions. According to Peierls, Landau, and Mermin, the long-range order in 2D should be destroyed by thermal fluctuations [91–94]. This analysis considered truly 2D material without defects, but not a 2D system which is a part of larger 3D structure. In this case, stability of a 2D crystal can be supported by a substrate or existing disorder (crumpling). On the other hand, graphene suspended above a substrate was demonstrated in 2007 [62]. These graphene membranes were stable under ambient conditions. It was shown by transmission electron microscopy (TEM) that graphene had high-quality lattice with occasional point defects [95]. Stability was enabled through elastic deformations in the third dimension related to interactions between bending and stretching long-wavelength phonons. The above conclusions were drawn from a nanobeam electron diffraction patterns which changed with the tilt angle. Diffraction peaks were sharp for normal incidence, but broadened for different angles, revealing that graphene is not perfectly flat. Samples were estimated to exhibit ripples with ~1 nm height and length of a few nanometers. It is expected that they can be created in a controllable way by thermally generated strains [96].

Experiments on graphene membranes allowed to estimate rigidity, elasticity and thermal conductivity. Lee et al. and Bunch et al. performed experiments and numerical simulations on graphene strength and elasticity [97, 98]. They determined an intrinsic strength which is the maximum pressure that can be supported by the defect-free material. Obtained values correspond to the largest Young modulus ever measured, ~1 TPa. Such high value is responsible for graphene robustness and stiffness. It answers the question why large graphene membranes, with up to 100 μm, do not scroll or fold [99]. Additionally, results regarding elastic properties predict high tolerance against deformations, well beyond a linear regime [97]. Graphene also reveals high thermal conductivity, predicted by Mingo et al. [100] and measured by Balandin et al. [101]. The experiment required an unconventional technique of non-contact measurement, the confocal micro-Raman spectroscopy. Balandin et al. heated their sample with 488 nm laser light and observed a shift of Raman G peak with increasing excitation power. Experimental data were fitted to the equation for thermal conductivity due to acoustic phonons, giving a value at room temperature that exceeded 5,300 W/mK, almost twice the value found for carbon nanotubes.

2.4 Electronic Band Structure of Graphene

2.4.1 Tight-Binding Model

The electronic band structure of graphene was described by Wallace already in 1946 [3] and here we follow his derivation. A comparison of tight-binding model with results of ab-initio calculations can be found in Chap. 6 and in, e.g., [102].

We start with six electrons occupying the $1s^2$, $2s^2$, and $2p^2$ orbitals of carbon. The structural and electronic properties are dictated by the 4 valence electrons. Three of those valence electrons occupy the s, p_x and p_y orbitals and hybridize to form sp^2 bonds (sigma bonds) connecting neighboring atoms, as shown in Fig. 2.8. These hybridized orbitals are responsible for structural stability of graphene. The fourth valence electron occupies the p_z orbital orthogonal to the plane of graphene. The hybridization of p_z orbitals leads to the formation of Π bands in graphene. In the following, we will describe the electronic structure of graphene within the single p_z orbital tight-binding (TB) model [3]. The honeycomb lattice of graphene can be

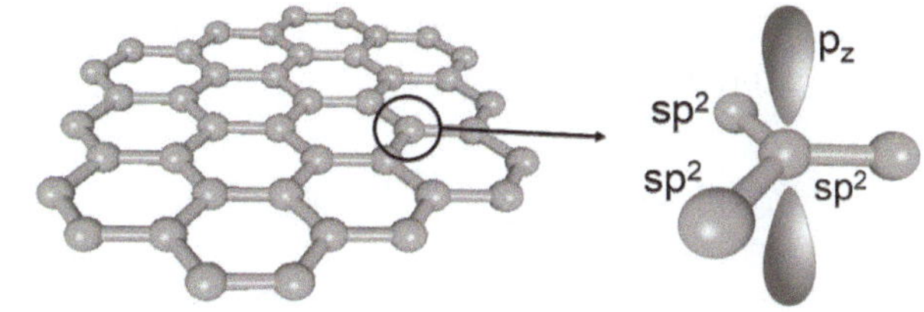

Fig. 2.8 A schematic plot of a graphene lattice (*left*) with atomic bonds (*right*) formed from valence electrons of a carbon atom. From four valence electrons, three on s, p_x and p_y orbitals form hybridized sp^2 bonds between neighboring lattice sites. The fourth valence electron occupies the p_z orbital orthogonal to the plane of graphene

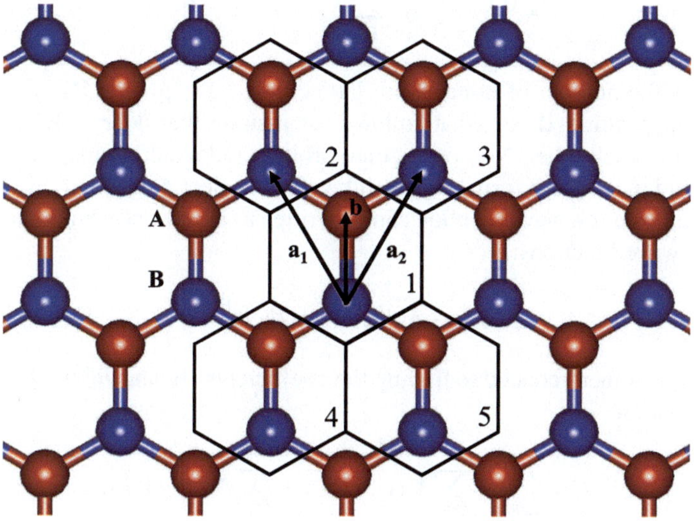

Fig. 2.9 Graphene honeycomb lattice. There are two atoms in a unit cell, A and B, distinguished by *red* and *blue* colors. Primitive unit vectors are defined as $\mathbf{a}_{1,2} = a/2(\pm\sqrt{3}, 3)$. $\mathbf{b} = a(0, 1)$ is a vector between two nearest neighboring atoms from the same unit cell

conveniently described in terms of two triangular Bravais sublattices represented with red and blue atoms in Fig. 2.9. The distance between nearest neighboring atoms is $b \approx 1.42$ Å. Primitive unit vectors can be defined as $\mathbf{a}_{1,2} = a/2(\pm\sqrt{3}, 3)$. Positions of all sublattice A and B atoms are then given by

$$\mathbf{R}_A = n\mathbf{a}_1 + m\mathbf{a}_2 + \mathbf{b}, \tag{2.1}$$

$$\mathbf{R}_B = n\mathbf{a}_1 + m\mathbf{a}_2, \tag{2.2}$$

where n and m are integers, and $\mathbf{b}$ is a vector going from the A atom to the B atom in a unit cell (see Fig. 2.9). There are two nonequivalent carbon atoms, A and B, in a unit cell.

The wave function of an electron on sublattice A can be written as a linear superposition of localized p_z orbitals of sublattice A:

$$\Psi_{\mathbf{k}}^{A}(\mathbf{r}) = \frac{1}{\sqrt{N_u}} \sum_{\mathbf{R}_A} e^{i\mathbf{k}\mathbf{R}_A} \phi_z(\mathbf{r} - \mathbf{R}_A). \tag{2.3}$$

Due to the translation symmetry and Bloch's theorem, the wave function is labeled by wave vector $\mathbf{k}$ and the coefficients of the expansion are given by $e^{i\mathbf{k}\mathbf{R}_A}$. The same applies to electron on the sublattice B:

$$\Psi_{\mathbf{k}}^{B}(\mathbf{r}) = \frac{1}{\sqrt{N_u}} \sum_{\mathbf{R}_B} e^{i\mathbf{k}\mathbf{R}_B} \phi_z(\mathbf{r} - \mathbf{R}_B). \tag{2.4}$$

Here N_u is the number of honeycomb lattice unit cells, $\phi_z(\mathbf{r} - \mathbf{R})$ is a p_z orbital localized at position $\mathbf{R}$. In what follows we assume that $\phi_z(\mathbf{r} - \mathbf{R})$ orbitals are orthogonal to each other. Non-orthogonal orbitals and resulting matrix elements of overlaps and the explicit form of ϕ_z will be given in Sect. 5.3.

The total electron wave function can be written as a linear combination of the two sublattice wave functions:

$$\Psi_{\mathbf{k}}(\mathbf{r}) = A_{\mathbf{k}} \Psi_{\mathbf{k}}^{A}(\mathbf{r}) + B_{\mathbf{k}} \Psi_{\mathbf{k}}^{B}(\mathbf{r}). \tag{2.5}$$

The problem is then reduced to finding the coefficients $A_{\mathbf{k}}$ and $B_{\mathbf{k}}$ by diagonalizing the Hamiltonian

$$H = \frac{p^2}{2m} + \sum_{\mathbf{R}_A} V(\mathbf{r} - \mathbf{R}_A) + \sum_{\mathbf{R}_B} V(\mathbf{r} - \mathbf{R}_B), \tag{2.6}$$

where $V(\mathbf{r} - \mathbf{R})$ is an effective atomic potential centered at $\mathbf{R}$. In other words, we need to calculate and diagonalize the matrix

$$H(\mathbf{k}) = \begin{pmatrix} \langle \Psi_{\mathbf{k}}^{A} | H | \Psi_{\mathbf{k}}^{A} \rangle & \langle \Psi_{\mathbf{k}}^{A} | H | \Psi_{\mathbf{k}}^{B} \rangle \\ \langle \Psi_{\mathbf{k}}^{B} | H | \Psi_{\mathbf{k}}^{A} \rangle & \langle \Psi_{\mathbf{k}}^{B} | H | \Psi_{\mathbf{k}}^{B} \rangle \end{pmatrix}, \tag{2.7}$$

with the assumption that $\Psi_{\mathbf{k}}^{A}$ and $\Psi_{\mathbf{k}}^{B}$ are orthogonal. Notice that we have

$$\left(\frac{p^2}{2m} + \sum_{\mathbf{R}_A} V(\mathbf{r} - \mathbf{R}_A) \right) \Psi_{\mathbf{k}}^{A} = \varepsilon_A(\mathbf{k}) \Psi_{\mathbf{k}}^{A}, \tag{2.8}$$

where, in the *nearest neighbor approximation*, $\varepsilon_A(\mathbf{k}) \approx 0$. This is due to the fact that the hopping integrals between neighboring sites on the same sublattice (i.e. next nearest neighbors in the honeycomb lattice) are neglected. Moreover, the constant onsite energies of p_z orbitals are taken to be zero. Next, we calculate $\langle \Psi_{\mathbf{k}}^{A} | H | \Psi_{\mathbf{k}}^{A} \rangle$:

$$\langle \Psi_{\mathbf{k}}^{A}|H|\Psi_{\mathbf{k}}^{A}\rangle = \frac{1}{N_u}\sum_{\mathbf{R}_A,\mathbf{R}'_A,\mathbf{R}_B} e^{i\mathbf{k}(\mathbf{R}_A-\mathbf{R}'_A)}\int d\mathbf{r}\phi_z^*(\mathbf{r}-\mathbf{R}'_A)V(\mathbf{r}-\mathbf{R}_B)\phi_z(\mathbf{r}-\mathbf{R}_A), \quad (2.9)$$

where the three-center integrals give zero in the nearest neighbor approximation. A similar result is obtained for $\langle \Psi_{\mathbf{k}}^{B}|H|\Psi_{\mathbf{k}}^{B}\rangle$. Thus, we have

$$\begin{aligned}\langle \Psi_{\mathbf{k}}^{A}|H|\Psi_{\mathbf{k}}^{A}\rangle &\approx 0,\\ \langle \Psi_{\mathbf{k}}^{B}|H|\Psi_{\mathbf{k}}^{B}\rangle &\approx 0.\end{aligned} \quad (2.10)$$

The off-diagonal term $\langle \Psi_{\mathbf{k}}^{B}|H|\Psi_{\mathbf{k}}^{A}\rangle$ gives

$$\langle \Psi_{\mathbf{k}}^{B}|H|\Psi_{\mathbf{k}}^{A}\rangle = \frac{1}{N_u}\sum_{\mathbf{R}_A,\mathbf{R}_B,\mathbf{R}'_B} e^{i\mathbf{k}(\mathbf{R}_A-\mathbf{R}_B)}\int d\mathbf{r}\phi_z^*(\mathbf{r}-\mathbf{R}_B)V(\mathbf{r}-\mathbf{R}'_B)\phi_z(\mathbf{r}-\mathbf{R}_A). \quad (2.11)$$

By neglecting three center integrals (taking $\mathbf{R}_B = \mathbf{R}'_B$), we obtain

$$\langle \Psi_{\mathbf{k}}^{B}|H|\Psi_{\mathbf{k}}^{A}\rangle = \frac{1}{N_u}\sum_{<\mathbf{R}_A,\mathbf{R}_B>} e^{i\mathbf{k}(\mathbf{R}_A-\mathbf{R}_B)}\int d\mathbf{r}\phi_z^*(\mathbf{r}-\mathbf{R}_B)V(\mathbf{r}-\mathbf{R}_B)\phi_z(\mathbf{r}-\mathbf{R}_A), \quad (2.12)$$

where the summation is now restricted to nearest neighbors only. The summation can be further expanded over three nearest neighbors as shown in Fig. 2.9. For a given pair of nearest neighbors at $\mathbf{R}_A$ and $\mathbf{R}_B$, the integral in the previous equation is a constant. This allows us to write

$$\begin{aligned}\langle \Psi_{\mathbf{k}}^{A}|H|\Psi_{\mathbf{k}}^{B}\rangle &= t\left(e^{-i\mathbf{k}\mathbf{b}} + e^{-i\mathbf{k}(\mathbf{b}-\mathbf{a}_1)} + e^{-i\mathbf{k}(\mathbf{b}-\mathbf{a}_2)}\right),\\ \langle \Psi_{\mathbf{k}}^{B}|H|\Psi_{\mathbf{k}}^{A}\rangle &= t\left(e^{i\mathbf{k}\mathbf{b}} + e^{i\mathbf{k}(\mathbf{b}-\mathbf{a}_1)} + e^{i\mathbf{k}(\mathbf{b}-\mathbf{a}_2)}\right),\end{aligned} \quad (2.13)$$

where we defined the hopping integral

$$t = \int d\mathbf{r}\phi_z^*(\mathbf{r}-\mathbf{R}_B)V(\mathbf{r}-\mathbf{R}_B)\phi_z(\mathbf{r}-\mathbf{R}_A), \quad (2.14)$$

for nearest neighbors $\mathbf{R}_A$ and $\mathbf{R}_B$. The value of t can be determined experimentally, and is usually taken to be $t \approx -2.8$ eV [103]. Finally, by defining

$$f(\mathbf{k}) = e^{-i\mathbf{k}\mathbf{b}} + e^{-i\mathbf{k}(\mathbf{b}-\mathbf{a}_1)} + e^{-i\mathbf{k}(\mathbf{b}-\mathbf{a}_1)}, \quad (2.15)$$

and using (2.7), (2.10), and (2.13), we can write the energy eigenequation system in the basis of A and B sublattice wave functions as

$$E(\mathbf{k})\begin{pmatrix} A_{\mathbf{k}} \\ B_{\mathbf{k}} \end{pmatrix} = t\begin{pmatrix} 0 & f(\mathbf{k}) \\ f^*(\mathbf{k}) & 0 \end{pmatrix}\begin{pmatrix} A_{\mathbf{k}} \\ B_{\mathbf{k}} \end{pmatrix}, \quad (2.16)$$

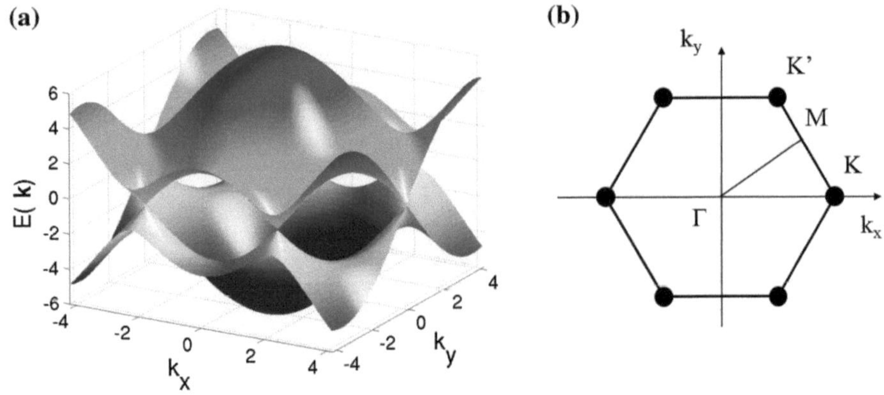

Fig. 2.10 **a** The band structure of graphene. The Fermi level is at $E(k) = 0$, where the valence and the conduction band touch each other in six points. These are corners of the first Brillouin zone, seen in a projection of the Brillouin zone shown in (**b**). From these six points only two are nonequivalent, indicated by K and K'. Other high symmetry points of reciprocal space are also indicated

whose solutions are

$$E_{\pm}(\mathbf{k}) = \pm|tf(\mathbf{k})| = \mp t|f(\mathbf{k})|,$$

corresponding to the conduction band with positive energy and the valence band with negative energy, plotted in Fig. 2.10. Using (2.3), (2.4), and (2.5), the corresponding conduction and valence band wave functions can be expressed as:

$$\Psi_{\mathbf{k}}^{c}(\mathbf{r}) = \frac{1}{\sqrt{2N_u}}\left(\sum_{\mathbf{R_A}} e^{i\mathbf{k}\mathbf{R_A}}\phi_z(\mathbf{r}-\mathbf{R_A}) - \sum_{\mathbf{R_B}} e^{i\mathbf{k}\mathbf{R_B}}\frac{f^*(\mathbf{k})}{|f(\mathbf{k})|}\phi_z(\mathbf{r}-\mathbf{R_B})\right),$$
$$\Psi_{\mathbf{k}}^{v}(\mathbf{r}) = \frac{1}{\sqrt{2N_u}}\left(\sum_{\mathbf{R_A}} e^{i\mathbf{k}\mathbf{R_A}}\phi_z(\mathbf{r}-\mathbf{R_A}) + \sum_{\mathbf{R_B}} e^{i\mathbf{k}\mathbf{R_B}}\frac{f^*(\mathbf{k})}{|f(\mathbf{k})|}\phi_z(\mathbf{r}-\mathbf{R_B})\right) \tag{2.17}$$

Note that the energy spectrum plotted in Fig. 2.10 is gapless at six K points in the Brillouin zone—graphene is a semimetal. The spectrum is symmetric around zero (Fermi level). This electron-hole symmetry is a consequence of retaining only nearest neighbor hopping; it is broken if one introduces a finite next-nearest neighbor hopping coupling similar to the one in (2.14). The behavior of charge carriers near the Fermi level has striking properties, as we will see in the next subsection.

2.4.2 Effective Mass Approximation, Dirac Fermions and Berry's Phase

For the charge-neutral system, each carbon atom gives one electron to the p_z orbital, for a total of $2N_u$ electrons in the honeycomb graphene lattice. As a result, the Fermi level is at $E(k) = 0$. From Fig. 2.10, it is seen that valence and conduction bands touch each other at six points. These are corners of the first Brillouin zone, also shown in the inset of the figure. Only two of these six points, indicated by K and K', are nonequivalent. The other four corners can be obtained by a translation by reciprocal vectors. In the inset, other high symmetry points of reciprocal space are also indicated, the Γ point in the center of the Brillouin zone and the M point. Here, we focus on low-energy electronic properties which correspond to states around K and K' points.

The conduction and valence energy dispersion $E(\mathbf{k})$ given by (2.16) can be expanded around $\mathbf{K}$ and $\mathbf{K}'$ points. Expansion of $f(\mathbf{k})$ around $\mathbf{K} = (4\pi/3\sqrt{3}a, 0)$ is given by

$$f(\mathbf{K}+\mathbf{q}) = f(\mathbf{K}) + f'(\mathbf{K})\mathbf{q} + \cdots, \tag{2.18}$$

where $\mathbf{q}$ is measured with respect to the $\mathbf{K}$ point. We get:

$$f(\mathbf{K}+\mathbf{q}) \approx -\frac{3}{2}a(q_x - iq_y). \tag{2.19}$$

(2.16) can then be written as

$$E_{\mathbf{K}}(\mathbf{q})\begin{pmatrix} A_{\mathbf{q}} \\ B_{\mathbf{q}} \end{pmatrix} = -\frac{3}{2}ta\begin{pmatrix} 0 & q_x - iq_y \\ q_x + iq_y & 0 \end{pmatrix}\begin{pmatrix} A_{\mathbf{q}} \\ B_{\mathbf{q}} \end{pmatrix}. \tag{2.20}$$

Eigenenergies can be found by diagonalizing the 2×2 matrix as before:

$$\begin{aligned} E^c_{\mathbf{K}}(\mathbf{q}) &= +\frac{3}{2}a|t||\mathbf{q}|, \\ E^v_{\mathbf{K}}(\mathbf{q}) &= -\frac{3}{2}a|t||\mathbf{q}|, \end{aligned} \tag{2.21}$$

and corresponding wave functions are given by

$$\begin{aligned} \Psi^c_{\mathbf{K}}(\mathbf{q}) &= \frac{1}{\sqrt{2}}\begin{pmatrix} e^{-i\theta_{\mathbf{q}}/2} \\ e^{+i\theta_{\mathbf{q}}/2} \end{pmatrix}, \\ \Psi^v_{\mathbf{K}}(\mathbf{q}) &= \frac{1}{\sqrt{2}}\begin{pmatrix} e^{-i\theta_{\mathbf{q}}/2} \\ -e^{+i\theta_{\mathbf{q}}/2} \end{pmatrix}, \end{aligned} \tag{2.22}$$

where we have defined $e^{i\theta_{\mathbf{q}}} = (q_x + iq_y)/|\mathbf{q}|$. In other words, $\theta_{\mathbf{q}}$ is defined as the angle of $\mathbf{q}$ measured from q_x-axis. Similar calculations can be done around the $\mathbf{K}'$ point. Of course, we obtain the same eigenenergies, but the eigenfunctions are now given by

$$\Psi^c_{\mathbf{K}'}(\mathbf{q}) = \frac{1}{\sqrt{2}}\begin{pmatrix} e^{+i\theta_{\mathbf{q}}/2} \\ e^{-i\theta_{\mathbf{q}}/2} \end{pmatrix},$$
$$\Psi^v_{\mathbf{K}'}(\mathbf{q}) = \frac{1}{\sqrt{2}}\begin{pmatrix} e^{+i\theta_{\mathbf{q}}/2} \\ -e^{-i\theta_{\mathbf{q}}/2} \end{pmatrix}. \tag{2.23}$$

Notice that, by introducing the Fermi velocity $v_F = 3|t|a/2\hbar$, and the Pauli matrix $\sigma = (\sigma_x, \sigma_y)$, the effective mass Hamiltonian in (2.20) can be rewritten as

$$H_K = -iv_F\sigma \cdot \nabla, \tag{2.24}$$

which is a 2D Dirac Hamiltonian acting on the two-component wavefunction $\Psi_{\mathbf{K}}$. The linear dispersion near K and K' points is thus strikingly different than the usual quadratic dispersion $q^2/2m$ for electrons with mass m. Instead, we have Dirac-like Hamiltonian for relativistic massless Fermions. Here, the role of the speed of light is played by the Fermi velocity. One can estimate $v_F \simeq 10^6\ m/s$ which is 300 times smaller than the speed of light in vacuum. Moreover, the eigenfunctions given in (2.22) consists of two components, in analogy with spinor wave functions for Fermions. Here, the role of the spin is played by two sublattices, A and B. These two-component eigenfunctions are called pseudospinors.

Let us now discuss the Berry's phase aspect of the pseudospinor. The energy spectra of the electron and hole form two Dirac cones touching at the Fermi level $E = 0$. This is an example of intersecting energy surfaces studied by Herzberg and Longuet-Higgins already in 1963 [104] and subsequently by Berry [105]. Let us consider the wave function of an electron with energy E on the upper section of Dirac cone propagating in the x direction. The wavevector is $q = q_x$, the angle $\theta_{\mathbf{q}}$ in (2.22) is $\theta_{\mathbf{q}} = 0$ and the wavefunction is explicitly given by:

$$\Psi^c_{\mathbf{K}}(\mathbf{q_x}) = \frac{1}{\sqrt{2}}\begin{pmatrix} 1 \\ 1 \end{pmatrix}.$$

If we now adiabatically move on the constant energy circle on the electron Dirac cone and return to the same direction of propagation $q = q_x$ we started with, the angle $\theta_{\mathbf{q}}$ in (2.22) is now $\theta_{\mathbf{q}} = 2\pi$. The new wavefunction now reads

$$\Psi^c_{\mathbf{K}}(\mathbf{q_x}*) = \frac{1}{\sqrt{2}}\begin{pmatrix} e^{-i2\pi/2} \\ e^{+i2\pi/2} \end{pmatrix} = \frac{1}{\sqrt{2}}\begin{pmatrix} e^{-i\pi} \\ e^{+i\pi} \end{pmatrix} = \frac{1}{\sqrt{2}}e^{-i\pi}\begin{pmatrix} 1 \\ 1 \end{pmatrix}.$$

We see that the wavefunction $\Psi^c_{\mathbf{K}}(\mathbf{q}^*_{\mathbf{x}})$ is the wavefunction we started with times the phase factor $e^{-i\pi}$, $\Psi^c_{\mathbf{K}}(\mathbf{q}^*_{\mathbf{x}}) = e^{-i\pi}\Psi^c_{\mathbf{K}}(\mathbf{q_x})$. The accumulated phase is the Berry's phase of Dirac electron in graphene.

2.4.3 Chirality and Absence of Backscattering

An important implication of pseudospin in graphene is the concept of chirality and absence of backscattering by impurity [106]. The chirality is related to the energy of a quasiparticle in the vicinity of the Dirac point, $H(\mathbf{k}) = \boldsymbol{\sigma} \cdot \mathbf{k}$. We see that for a constant energy the state k and $-k$ correspond to pseudospin σ and $-\sigma$. The electron propagating in the opposite direction must have the opposite pseudospin. To understand how pseudospin chirality affects backscattering, let us consider an impurity potential $V_{imp}(\mathbf{r})$ which is long ranged compared with the lattice constant, and smoothly varying over the unit cell. We would like to calculate the transition matrix element for a conduction electron from a state $\mathbf{q}$ to a state $\mathbf{q}'$:

$$\tau(\mathbf{q}, \mathbf{q}') = \langle \mathbf{q}'c|V_{imp}|\mathbf{q}c\rangle. \tag{2.25}$$

In the effective mass approximation, using (2.22) and (2.5), we get:

$$\begin{aligned}\tau(\mathbf{q}, \mathbf{q}') = \frac{1}{2N_u}\int d^2r & \left(e^{-i\theta_{\mathbf{q}'}/2}\sum_{\mathbf{R_A}} e^{-i(\mathbf{K}+\mathbf{q}')\mathbf{R_A}}\phi_z(\mathbf{r}-\mathbf{R_A})\right.\\ & \left.+ e^{+i\theta_{\mathbf{q}'}/2}\sum_{\mathbf{R_B}} e^{-i(\mathbf{K}+\mathbf{q}')\mathbf{R_B}}\phi_z(\mathbf{r}-\mathbf{R_B})\right)\\ \times V_{imp}(\mathbf{r}) & \left(e^{+i\theta_{\mathbf{q}}/2}\sum_{\mathbf{R_A}} e^{+i(\mathbf{K}+\mathbf{q})\mathbf{R_A}}\phi_z(\mathbf{r}-\mathbf{R_A})\right.\\ & \left.+ e^{-i\theta_{\mathbf{q}}/2}\sum_{\mathbf{R_B}} e^{+i(\mathbf{K}+\mathbf{q})\mathbf{R_B}}\phi_z(\mathbf{r}-\mathbf{R_B})\right),\end{aligned} \tag{2.26}$$

where we ignored complex conjugation of ϕ_z orbitals since they are taken to be real. Two of the four integrals are of the type:

$$\int d^2r\phi_z(\mathbf{r}-\mathbf{R_1})V_{imp}(\mathbf{r})\phi_z(\mathbf{r}-\mathbf{R_2}) \approx V_{imp}(\mathbf{R_1})\delta(\mathbf{R_1}-\mathbf{R_2}) \tag{2.27}$$

since (i) for nearest neighbors $V_{imp}(\mathbf{r})$ is a smoothly varying function over the unit cell and can be taken out of the integral, (ii) orbitals have zero overlap if they are far away from each other. This leaves us with

$$\begin{aligned}\tau(\mathbf{q}, \mathbf{q}') = \frac{1}{2N_u} & \left(e^{-i\Delta\theta/2}\sum_{\mathbf{R_A}} e^{-i(\mathbf{q}+\mathbf{q}')\mathbf{R_A}}V_{imp}(\mathbf{R_A})\right.\\ & \left.+ e^{+i\Delta\theta/2}\sum_{\mathbf{R_B}} e^{-i(\mathbf{q}+\mathbf{q}')\mathbf{R_B}}V_{imp}(\mathbf{R_B})\right),\end{aligned}$$

where $\Delta\theta = \theta_{\mathbf{q}'} - \theta_{\mathbf{q}}$, i.e. the angle between the incoming wave and scattered wave. The two terms represent scattering matrix elements of the A and B sublattice components of the pseudospinor. The two summations present in each term represent the Fourier transform of V_{imp} over A and B sublattices. They are equal in the continuum limit for a long-ranged and smoothly varying V_{imp}. Thus, we have

$$\tau(\mathbf{q}, \mathbf{q}') = cos(\Delta\theta/2) F_{\mathbf{q}+\mathbf{q}'}\{V_{imp}\}. \tag{2.28}$$

Clearly, as $\Delta\theta$ approaches π, i.e. for a backscattering event, the transition element $\tau(\mathbf{q}, \mathbf{q}')$ vanishes. This destructive interference between the sublattices leads to the absence of backscattering, and is responsible of high conductivity of graphene. A more general proof of the absence of backscattering in graphene can be found in [106].

2.4.4 Bilayer Graphene

The tight-binding model discussed in Sect. 2.4.1 can also be generalized to bilayer graphene [14, 21, 23]. Starting with two degenerate Dirac cones the interlayer tunneling leads to splitting off of the two bands, while the remaining two conduction and valence bands touch at the Fermi level. The quasiparticles have a finite mass but there is no gap, as shown in Fig. 2.11. One of the most interesting aspects of bilayer graphene is the possibility to open a gap in the energy spectrum by applying an external electric field perpendicular to the layers [107–113]. In this section, following our earlier work [14, 23], we demonstrate the opening of the gap as a function of potential difference between the layers due to an applied perpendicular electric field. In Sect. 2.4.1 we showed that a graphene layer is described by a linear combination of two sublattice wave functions $\Psi_{\mathbf{k}}^{A}(\mathbf{r})$ and $\Psi_{\mathbf{k}}^{B}(\mathbf{r})$. In the bilayer case, we now have four wave functions corresponding to A_1 and B_1 sublattices in the first layer and A_2 and B_2 sublattices in the second layer (see Fig. 2.11):

$$\Psi_{\mathbf{k}}^{A_1}(\mathbf{r}) = \frac{1}{\sqrt{N_u}} \sum_{\mathbf{R}_{A_1}} e^{i\mathbf{k}\mathbf{R}_{A1}} \phi_z(\mathbf{r} - \mathbf{R}_{A_1}), \tag{2.29}$$

$$\Psi_{\mathbf{k}}^{B_1}(\mathbf{r}) = \frac{1}{\sqrt{N_u}} \sum_{\mathbf{R}_{B_1}} e^{i\mathbf{k}\mathbf{R}_{B1}} \phi_z(\mathbf{r} - \mathbf{R}_{B_1}), \tag{2.30}$$

$$\Psi_{\mathbf{k}}^{A_2}(\mathbf{r}) = \frac{1}{\sqrt{N_u}} \sum_{\mathbf{R}_{A_2}} e^{i\mathbf{k}\mathbf{R}_{A2}} \phi_z(\mathbf{r} - \mathbf{R}_{A_2}), \tag{2.31}$$

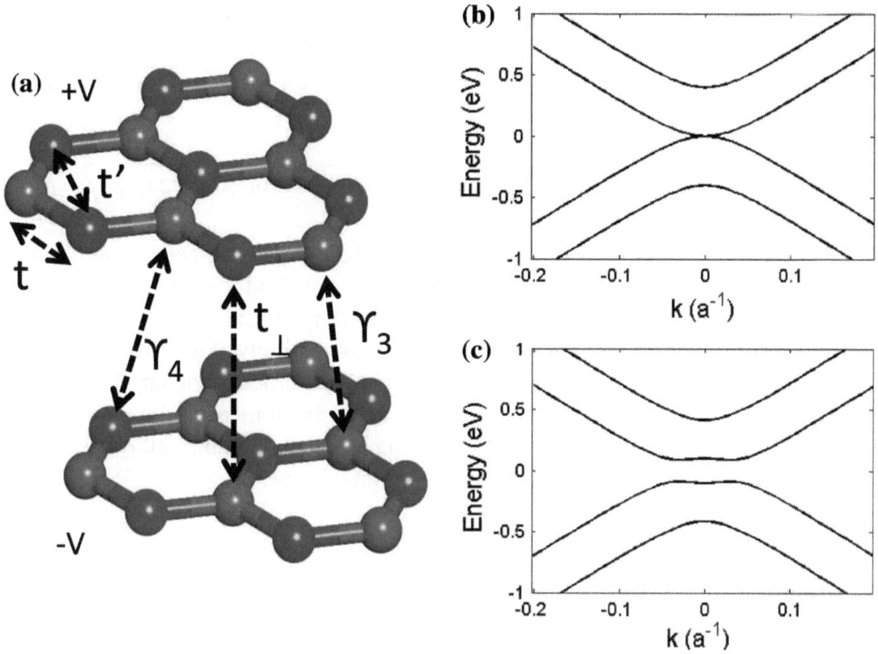

Fig. 2.11 **a** A schematic plot of tight-binding parameters in bilayer graphene and **b** energy spectra in the absence (*upper*) and in the presence (*lower*) of electric field

$$\Psi_{\mathbf{k}}^{B_2}(\mathbf{r}) = \frac{1}{\sqrt{N_u}} \sum_{\mathbf{R}_{B_2}} e^{i\mathbf{k}\mathbf{R}_{B_2}} \phi_z(\mathbf{r} - \mathbf{R}_{B_2}). \tag{2.32}$$

We now need to describe the hopping parameters between atoms in different layers. In Fig. 2.11a we show two layers arranged in the AB stacking of 3D graphite, also called Bernal stacking [12, 14, 109]. In such situation, the A_2 sublattice in the upper layer is directly above the B_1 sublattice of the lower sublattice. Thus, the strongest inter-layer hopping elements occur between the A_2 atoms and B_1 atoms, described by the parameter $t_\perp$. Other relevant inter-layer hopping parameters are commonly denoted as γ_3 between B_2 atoms and B_1 atoms, and γ_4 between B_2 atoms and A_1 atoms, both weaker than $t_\perp$. For graphite, values of inter-layer hopping elements are given by $t_\perp \approx -0.4$ eV, $\gamma_3 \approx -0.04$ eV, and $\gamma_4 \approx -0.3$ eV. For simplicity, in the following we will take $\gamma_3 = \gamma_4 = 0$.

It is then possible to write an effective Hamiltonian around a K-point similar to 2.20

$$E(\mathbf{k}) \begin{pmatrix} A_{1\mathbf{k}} \\ B_{1\mathbf{k}} \\ A_{2\mathbf{k}} \\ B_{2\mathbf{k}} \end{pmatrix} = - \begin{pmatrix} -V & \frac{3}{2}ta\mathbf{k}^* & 0 & 0 \\ \frac{3}{2}ta\mathbf{k} & -V & t_\perp & 0 \\ 0 & t_\perp & V & \frac{3}{2}ta\mathbf{k}^* \\ 0 & 0 & \frac{3}{2}ta\mathbf{k} & V \end{pmatrix} \begin{pmatrix} A_{1\mathbf{k}} \\ B_{1\mathbf{k}} \\ A_{2\mathbf{k}} \\ B_{2\mathbf{k}} \end{pmatrix}, \tag{2.33}$$

where we now have a four-component spinor instead of two. We have also added a potential difference of $2V$ between the two layers to model the effect of applied electric field. The above four-by-four matrix can be solved exactly using standard techniques to give

$$E_{\pm}^{2}(\mathbf{k}) = V^2 + 9t^2a^2k^2/4 + t_{\perp}^2/2 \pm \sqrt{9V^2t^2a^2k^2 + 9t^4a^2k^2/4 + t_{\perp}^4/4}. \quad (2.34)$$

In Fig. 2.11b, c we plot the energy spectrum of the bilayer graphene using 2.34 for V = 0 and V = 0.1 eV respectively. For V = 0 we see that the dispersion relation is no more linear but parabolic as can also be deduced from 2.34. However, the energy gap is still zero giving a metallic behavior. Most interestingly, if a small electric field is applied, i.e. for nonzero V, there opens a gap of the order of the applied bias 2 V. The dependence of the gap on the applied bias has been measured experimentally [108, 110–113]. The tunability of the gap with electric field makes bilayer graphene interesting from a technological application point of view.

References

1. S. Ijima, Nature **354**(6348), 5658 (1991)
2. H.W. Kroto, J.R. Heath, S.C. O'Brien, R.F. Curl, R.E. Smalley, Nature **318**, 162–164 (1985)
3. P.R. Wallace, Phys. Rev. **71**, 622 (1947)
4. P.A. Dirac, Proc. R. Soc. Lond. A **117**, 610–624 (1928)
5. H. Weyl, Proc. Natl. Acad. Sci. **15**, 323 (1929)
6. S.Y. Zhou, G.-H. Gweon, J. Graf, A.V. Fedorov, C.D. Spataru, R.D. Diehl, Y. Kopelevich, D.-H. Lee, S.G. Louie, A. Lanzara, Nature Phys. **2**, 595–599 (2006)
7. J.W. McClure, Phys. Rev. **108**, 612 (1957)
8. J.C. Slonczewski, P.R. Weiss, Phys. Rev. **109**, 272 (1958)
9. G. Dresselhaus, M.S. Dresselhaus, Phys. Rev. **140**, A401–A412 (1965)
10. G.W. Semenoff, Phys. Rev. Lett. **53**, 2449 (1984)
11. D. Haldane, Phys. Rev. Lett. **61**, 2015 (1988)
12. M.S. Dresselhaus, G. Dresselhaus, Adv. Phys. **30**, 139 (1981)
13. E.J. Mele, J.J. Ritsko, Phys. Rev. Lett. **43**, 68 (1979)
14. D.M. Hoffman, P.C. Eklund, R.E. Heinz, P.Hawrylak, K.R. Subbaswamy, Phys.Rev. B **31** 3973 (1984)
15. J. Kouvetakis, R.B. Kaner, M.L. Sattler, N. Bartlett, J. Chem. Soc. Chem. Commun. **1986**, 1758 (1986)
16. P. Hawrylak, K.R. Subbaswamy, Phys. Rev. Lett. **53**, 2098–2101 (1984)
17. G. Kirczenow, Phys. Rev. Lett. **55**, 2810 (1985)
18. M.S. Dresselhaus, G. Dresselhaus, J.E. Fisher, Phys. Rev. B **15**, 3180 (1977)
19. R.C. Tatar, S. Rabii, Phys. Rev. B **25**, 4126 (1982)
20. D.P. DiVincenzo, E.J. Mele, Phys. Rev. B **29**, 1685 (1984)
21. J. Blinowski, N.H. Hau, C. Rigaux, J.P. Vieren, R. Le Toullec, G. Furdin, A. Herold, J. Melin, J. Phys. (Paris) **41**, 47 (1980)
22. W.-K. Kenneth Shung. Phys. Rev. B **34**, 979 (1986)
23. P. Hawrylak, Solid State Commun. **63**, 241 (1987)
24. K.W.-K. Shung, G.D. Mahan, Phys. Rev. B **38**, 3856 (1988)
25. H.P. Boehm, A. Clauss, G.O. Fischer, U. Hofmann, Anorg. Allg. Chem. **316**, 119 (1962)
26. H.P. Boehm, R. Setton, E. Stumpp, Carbon **24**, 241 (1986)

27. J.T. Grant, T.W. Haas, Surf. Sci. **21**, 76 (1970)
28. J.M. Blakely, J.S. Kim, H.C. Potter, J. Appl. Phys. **41**, 2693 (1970)
29. A.J. van Bommel, J.E. Crombeen, A. van Tooren, Surf. Sci. **48**, 463 (1975)
30. A. Nagashima, K. Nuka, K. Satoh, H. Itoh, T. Ichinokawa, C. Oshima, S. Otani, Surf. Sci. **287–288**, 609 (1993)
31. K. Seibert, G.C. Cho, W. Kütt, H. Kurz, D.H. Reitze, J.I. Dadap, H. Ahn, M.C. Downer, A.M. Malvezzi, Phys. Rev. B **42**, 2842 (1990)
32. Y. Ohashi, T. Koizumi, T. Yoshikawa, T. Hironaka, K. Shiiki, TANSO **180**, 235 (1997)
33. M. Müller, C. Kbel, K. Müllen, Chem. Eur. J. **4**, 2099 (1998)
34. N. Tyutyulkov, G. Madjarova, F. Dietz, K. Müllen, J. Phys. Chem. B **102**, 10183 (1998)
35. X. Lu, M. Yu, H. Huang, R. S Ruoff. Nanotechnology **10**, 269 (1999)
36. K.S. Novoselov, A.K. Geim, S.V. Morozov, D. Jiang, Y. Zhang, S.V. Dubonos, I.V. Grigorieva, A.A. Firsov, Science **306**, 666 (2004)
37. C. Berger, Z. Song, T. Li, X. Li, A.Y. Ogbazghi, R. Feng, Z. Dai, A.N. Marchenkov, E.H. Conrad, P.N. First, W.A. de Heer, J. Phys. Chem. B **108**, 19912 (2004)
38. A.C. Ferrari, J.C. Meyer, V. Scardaci, C. Casiraghi, M. Lazzeri, F. Mauri, S. Piscanec, D. Jiang, K.S. Novoselov, S. Roth, A.K. Geim, Phys. Rev. lett. **97**, 187401 (2006)
39. K.S. Novoselov, A.K. Geim, S.V. Morozov, D. Jiang, M.I. Katsnelson, I.V. Grigorieva, S.V. Dubonos, A.A. Firsov, Nature **438**, 197 (2005)
40. A.K. Geim, Rev. Mod. Phys. **83**, 851 (2011)
41. Y. Zhang, Y.W. Tan, H.L. Stormer, P. Kim, Nature **438**, 201 (2005)
42. K.S. Novoselov, Z. Jiang, Y. Zhang, S.V. Morozov, H.L. Stormer, U. Zeitler, J.C. Maan, G.S. Boebinger, P. Kim, A.K. Geim, Science **315**, 1379 (2007)
43. K.S. Novoselov, A.K. Geim, Nature **6**, 183 (2007)
44. M.L. Sadowski, G. Martinez, M. Potemski, C. Berger, W.A. de Heer, Phys. Rev. Lett. **97**, 266405 (2006)
45. M.I. Katsnelson, K.S. Novoselov, A.K. Geim, Nature Phys. **2**, 620 (2006)
46. M.I. Katsnelson, Materials Today **10**, 20–27 (2007)
47. A.F. Young, P. Kim, Nature Phys. **5**, 222 (2009)
48. N. Stander, B. Huard, D. Goldhaber-Gordon, Phys. Rev. Lett. **102**, 026807 (2009)
49. V.N. Kotov, B. Uchoa, V.M. Pereira, F. Guinea, A.H. Castro, Neto. Rev. Mod. Phys. **84**, 10671125 (2012)
50. S. Sorella, E. Tosatti, Europhys. Lett. **19**, 699 (1992)
51. S. Sorella, Y. Otsuka, S. Yunoki, Scientific Reports **2**, 992 (2012)
52. F. Wang, Y.B. Zhang, C.S. Tian, C. Girit, A. Zettl, M. Crommie, Y.R. Shen, Science **320**, 206 (2008)
53. K.F. Mak, M.Y. Sfeir, Y. Wu, C.H. Lui, J.A. Misewich, T.F. Heinz, Phys. Rev. Lett. **101**, 196405 (2008)
54. R.R. Nair, P. Blake, A.N. Grigorenko, K.S. Novoselov, T.J. Booth, T. Stauber, N.M.R. Peres, A.K. Geim, Science **320**, 1308 (2008)
55. P. Avouris, F. Xia, MRS Bulletin **37**, 1225 (2012)
56. L. Yang, J. Deslippe, C.-H. Park, M.L. Cohen, S.G. Louie, Phys. Rev. Lett. **103**, 186802 (2009)
57. Y. Lin, K.A. Jenkins, A. Valdes-Garcia, J.P. Small, D.B. Farmer, P. Avouris, Nano Lett. **9**, 422 (2009)
58. K.S. Kim, Y. Zhao, H. Jang, S.Y. Lee, J.M. Kim, K.S. Kim, J.-H. Ahn, P. Kim, J.-Y. Choi, B.H. Hong, Nature **457**, 706 (2009)
59. F. Xia, T. Mueller, Y. Lin, A. Valdes-Garcia, P. Avouris, Nat. Nanotechnol. **4**, 839 (2009)
60. T. Mueller, F. Xia, P. Avouris, Nature Photon. **4**, 297 (2010)
61. T. Ihn, J. Gttinger, F. Molitor, S. Schnez, E. Schurtenberger, A. Jacobsen, S. Hellmller, T. Frey, S. Drscher, C. Stampfer, K. Ensslin, Materials Today **44**, 20–27 (2010)
62. J.C. Meyer, A.K. Geim, M.I. Katsnelson, K.S. Novoselov, T.J. Booth, S. Roth, Nature **446**, 60 (2007)

63. K.I. Bolotin, K.J. Sikes, Z. Jiang, M. Klimac, G. Fudenberg, J. Hone, P. Kim, H.L. Stormer, Solid State Commun. **146**, 351 (2008)
64. K.I. Bolotin, K.J. Sikes, J. Hone, H.L. Stormer, P. Kim, Phys. Rev. Lett. **101**, 096802 (2008)
65. S.V. Morozov, K.S. Novoselov, M.I. Katsnelson, F. Schedin, D.C. Elias, J.A. Jaszczak, A.K. Geim, Phys. Rev. Lett. **100**, 016602 (2008)
66. S. Bae, H. Kim, Y. Lee, X. Xu, J.-S. Park, Y. Zheng, J. Balakrishnan, T. Lei, H.R. Kim, Y.I. Song, Y.-J. Kim, K.S. Kim, B. Özyilmaz, J.-H. Ahn, B.H. Hong, S. Iijima, Nat. Nanotechnol. **5**, 574 (2010)
67. B. Trauzettel, D.V. Bulaev, D. Loss, G. Burkard, Nature **3**, 192 (2007)
68. A. Rycerz, J. Tworzydlo, C.W.J. Beenakker, Nature Phys. **3**, 172 (2007)
69. F. Schedin, A.K. Geim, S.V. Morozov, E.W. Hill, P. Blake, M.I. Katsnelson, K.S. Novoselov, Nature **6**, 652 (2007)
70. S. Garaj, W. Hubbard, A. Reina, J. Kong, D. Branton, J.A. Golovchenko, Nature **467**, 190 (2010)
71. V. Yu, E. Whiteway, J. Maassen, M. Hilke, Phys. Rev. B **84**, 205407 (2011)
72. A. Reina, X. Jia, J. Ho, D. Nezich, H. Son, V. Bulovic, M.S. Dresselhaus, J. Kong, Nano Lett. **9**, 30 (2009)
73. X. Li, W. Cai, J. An, S. Kim, J. Nah, D. Yang, R. Piner, A. Velamakanni, I. Jung, E. Tutuc, S.K. Banerjee, L. Colombo, R.S. Ruoff, Science **324**, 1312 (2009)
74. Z. Yan, J. Lin, Z. Peng, Z. Sun, Y. Zhu, L. Li, C. Xiang, E.L. Samuel, C. Kittrell, J.M. Tour, ACS Nano **6**, 9110 (2012)
75. C. Berger, Z. Song, X. Li, X. Wu, N. Brown, C. Naud, D. Mayou, T. Li, J. Hass, A.N. Marchenkov, E.H. Conrad, P.N. First, W.A. de Heer, Science **312**, 1191 (2006)
76. J. Hass, R. Feng, T. Li, X. Li, Z. Zong, W.A. de Heer, P.N. First, E.H. Conrada, C.A. Jeffrey, C. Berger, Appl. Phys. Lett. **89**, 143106 (2006)
77. J. Hass, F. Varchon, J.E. Millán-Otoya, M. Sprinkle, N. Sharma, W.A. de Heer, C. Berger, P.N. First, L. Magaud, E.H. Conrad, Phys. Rev. Lett. **100**, 125504 (2008)
78. J. Hass, W.A. de Heer, E.H. Conrad, J. Phys.: Condens. Matter **20**, 323202 (2008)
79. K.V. Emtsev, A. Bostwick, K. Horn, J. Jobst, G.L. Kellogg, L. Ley, J.L. McChesney, T. Ohta, S.A. Reshanov, J. Rhrl, E. Rotenberg, A.K. Schmid, D. Waldmann, H.B. Weber, T. Seyller, Nature Mater. **8**, 203 (2009)
80. J. Hicks, A. Tejeda, A.A. Taleb-Ibrahimi, M.S.M.S. Nevius, F.F. Wang, K.K. Shepperd, J.J. Palmer, F. Bertran, P. Le Fvre, J. Kunc, W.A. de Heer, C. Berger, E.H. Conrad, A wide band gap metal-semiconductor-metal nanostructure made entirely from graphene. Nature Phys. **9**, 49 (2013)
81. J. Borysiuk, R. Bożek, W. Strupiński, A. Wysmołek, K. Grodecki, R. Stepniewski, J.M. Baranowski, J. Appl. Phys. **105**, 023503 (2009)
82. J. Krupka, W. Strupiński, Appl. Phys. Lett. **96**, 082101 (2010)
83. W. Strupiński, K. Grodecki, A. Wysmołek, R. Stepniewski, T. Szkopek, P.E. Gaskell, A. Grüneis, D. Haberer, R. Bożek, J. Krupka, J.M. Baranowski, Nano Lett. **11**, 1786 (2011)
84. W.S. Hummers, R.E. Offeman, J. Am. Chem. Soc. **80**, 1339 (1958)
85. S. Stankovich, D.A. Dikin, G.H.B. Dommett, K.M. Kohlhaas, E.J. Zimney, E.A. Stach, R.D. Piner, S.T. Nguyen, R.S. Ruoff, Nature **442**, 282 (2006)
86. D.A. Dikin, S. Stankovich, E.J. Zimney, R.D. Piner, G.H.B. Dommett, G. Evmenenko, S.T. Nguyen, R.S. Ruoff, Nature **448**, 457 (2007)
87. M.J. McAllister, J. Li, D.H. Adamson, H.C. Schniepp, A.A. Abdala, J. Liu, M. Herrera-Alonso, D.L. Milius, R. Car, R.K. Prudhomme, I.A. Aksay, Chem. Mater. **19**, 4396 (2007)
88. S. Stankovich, D.A. Dikin, R.D. Piner, K.A. Kohlhaas, A. Kleinhammes, Y. Jia, Y. Wu, S.T. Nguyen, R.S. Ruoff, Carbon **45**, 1558 (2007)
89. I. Jung, D.A. Dikin, R.D. Piner, R.S. Ruoff, Nano Lett. **8**, 4283 (2008)
90. K.S. Novoselov, D. Jiang, F. Schedin, T.J. Booth, V.V. Khotkevich, S.V. Morozov, A.K. Geim, PNAS **102**, 10451 (2005)
91. R.E. Peierls, Helv. Phys. Acta **7**, 81 (1934)
92. R.E. Peierls, Ann. Inst. H. Poincare **5**, 177 (1935)

93. L.D. Landau, E.M. Lifshitz, *Statistical Physics* (Part I, Pergamon, Oxford, UK, 1980)
94. N.D. Mermin, Phys. Rev. **176**, 250 (1968)
95. J.C. Meyer, C. Kisielowski, R. Erni, M.D. Rossell, M.F. Crommie, A. Zettl, Nano Lett. **8**, 3582 (2008)
96. W. Bao, F. Miao, Z. Chen, H. Zhang, W. Jang, C. Dames, C.N. Lau, Nature Nanotech. **4**, 562 (2007)
97. C. Lee, X. Wei, J.W. Kysar, J. Hone, Science **321**, 385 (2008)
98. J.S. Bunch, S.S. Verbridge, J.S. Alden, A.M. van der Zande, J.M. Parpia, H.G. Craighead, P.L. McEuen, Nano Lett. **8**, 2458 (2008)
99. T.J. Booth, P. Blake, R.R. Nair, D. Jiang, E.W. Hill, U. Bangert, A. Bleloch, M. Gass, K.S. Novoselov, M.I. Katsnelson, A.K. Geim, Nano Lett. **8**, 2442 (2008)
100. N. Mingo, D.A. Broido, Phys. Rev. Lett. **95**, 096105 (2005)
101. A.A. Balandin, S. Ghosh, W. Bao, I. Calizo, D. Teweldebrhan, F. Miao, C.N. Lau, Nano Lett. **8**, 902 (2008)
102. S. Reich, J. Maultzsch, C. Thomsen, P. Ordejón, Phys. Rev. B **66**, 035412 (2002)
103. A.H.C. Neto, F. Guinea, N.M.R. Peres, K.S. Novoselov, A.K. Geim, Rev. Mod. Phys. **81**, 109 (2009)
104. G. Herzberg, H.C. Longuet-Higgins, Discuss. Faraday Soc. **35**, 77 (1963)
105. M.V. Berry, Proc. R. Soc. Lond. A **392**, 45 (1984)
106. T. Ando, Y. Zheng, H. Suzuura, J. Phys. Soc. Jpn. **71**, 1318 (2002)
107. E. McCann, V.I. Falko, Phys. Rev. Lett. **96**, 086805 (2006)
108. T. Ohta, A. Bostwick, T. Seyller, K. Horn, E. Rotenberg, Science **313**, 951 (2006)
109. E. McCann, D.S.L. Abergel, V.I. Fal'ko, Solid State Commun. **143**, 110 (2007)
110. E.V. Castro, K.S. Novoselov, S.V. Morozov, N.M.R. Peres, J.M.B. Lopes dos Santos, J. Nilsson, F. Guinea, A.K. Geim, A.H. Castro, Neto. Phys. Rev. Lett. **99**, 216802 (2007)
111. J.B. Oostinga, H.B. Heersche, X. Liu, A.F. Morpurgo, L.M.K. Vandersypen, Nat. Mater. **7**, 151 (2008)
112. K.F. Mak, C.H. Lui, J. Shan, T.F. Heinz, Phys. Rev. Lett. **102**, 256405 (2009)
113. Y. Zhang, T.T. Tang, C. Girit, Z. Hao, M.A. Martin, A. Zettl, M.F. Crommie, Y.R. Shen, F. Wang, Nature **459**, 820 (2009)

Chapter 3
Graphene Nanostructures and Quantum Dots

Abstract This chapter describes the fabrication methods and experiments on graphene nanostructures and quantum dots, with focus on the role of edges and size quantization effects.

Considerable interest in graphene is related to potential electronic applications, e.g., as transistors, transparent electrodes or photodetectors [1]. In the case of, e.g., switching transistor on and off, energy gap is needed to control the current. However, since graphene is a semiconductor with a zero-energy band gap and a minimum conductivity at the Dirac point, the current cannot be switched off. Additionally, as a result of the Klein paradox, it is difficult to confine electrons by an electrostatic gate. The problem of zero-energy gap can be solved by reducing the lateral size of graphene. As a result of size quantization, an energy gap opens. Finite-size semi-metallic graphene becomes a semiconductor. Among graphene nanostructures, graphene ribbons (strips) and graphene quantum dots (islands) are of particular interest. Cutting graphene nanostructures out of graphene results in two types of edges, armchair and zigzag, as illustrated in Fig. 3.1. The graphene nanostructure can also be characterized by whether the sublattice symmetry is conserved or not. As we will show, both types of edge and presence or absence of sublattice symmetry play an important role in determining electronic properties of graphene nanostructures.

3.1 Fabrication Methods

Graphene can be patterned into ribbons (GNR) with different widths by use of electron-beam lithography and an etching mask, as proposed by, e.g., P. Kim's group [2, 3]. One starts from high-quality graphene obtained by mechanical exfoliation. Next, graphene is deposited onto heavily p-doped Si substrate covered by SiO_2 layer. Strips of graphene are covered by a protective etch mask made with cubical-shaped molecules having one Si atom at each corner, with corners being linked via oxygen atoms, hydrogen forming silsesquioxane (HSQ). The unprotected graphene is etched away by the oxygen plasma. By using this technique, Kim's group was able to perform transport measurements on samples with widths from 20 to 500 nm and lengths $\sim$1 μm. They noted that transport properties strongly depend on both boundary scattering and trapped charges in the substrate.

A.D. Güçlü et al., *Graphene Quantum Dots*,
NanoScience and Technology, DOI 10.1007/978-3-662-44611-9_3

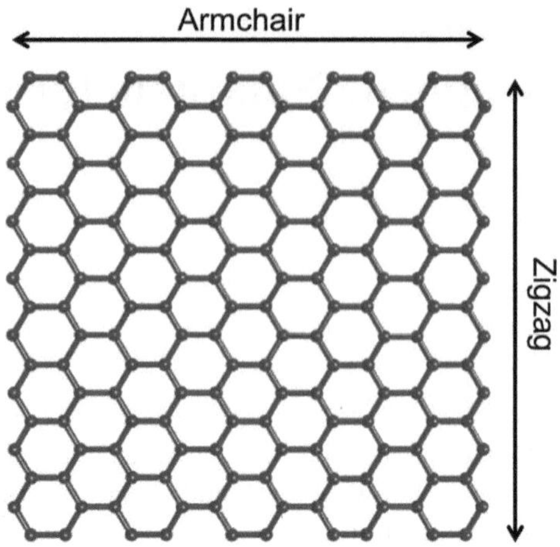

Fig. 3.1 Schematic illustration of two possible edge termination of graphene quantum dot

A different method of creating ribbons was proposed by Jia et al. [4–6]. They used Joule heating and electron beam irradiation [4]. Samples were exposed to electron irradiation for 20 min. and heated by directional high electrical current. During the heating, carbon atoms on sharp edges evaporated and GNRs with smooth edges were created.

Li et al. chemically derived graphene nanoribbons with well-defined edges [7]. The width of ribbons varied from $\sim$10 to 50 nm with length $\sim$1 μm. Graphene nanostructures with irregular shapes were also reported. They observed ribbons with 120° kink and zigzag edges. While the above work studied the thinnest ribbons with $\sim$10 nm width, Cai et al. proposed a method of creating ribbons with width less than $\sim$1 nm [8]. They started from colligated monomers, which define the width of the ribbon. These monomers were deposited onto the clean substrate surfaces by sublimation from a sixfold evaporator. They used two-step annealing process with different temperatures for straight and so-called chevron-type ribbons. Many other chemical approaches to create graphene quantum nanostructures with different shapes were also proposed [9–13]. Different shapes imply different chirality of the graphene nanoribbon. Chirality is related to the angle at which a ribbon is cut. GNRs, having different chiralities and widths, were chemically synthesized by unzipping a carbon nanotube [14, 15]. The presence of 1D GNR edge states was confirmed by using STM. The comparison of experimental results with the theoretical prediction based on the Hubbard model and density functional theory (DFT) calculations provided an evidence for the formation of spin-polarized edge states [15–18]. It was shown that electronic and magnetic properties can be tuned by changing the edge chirality and the width [19]. Partially unzipped carbon nanotubes were also studied [20, 21]. Topological defects similar to that at the interface between two graphene layers were considered. An appearance of spatially localized interface states was predicted [20] and general rules for the existence of edge states were discussed [22].

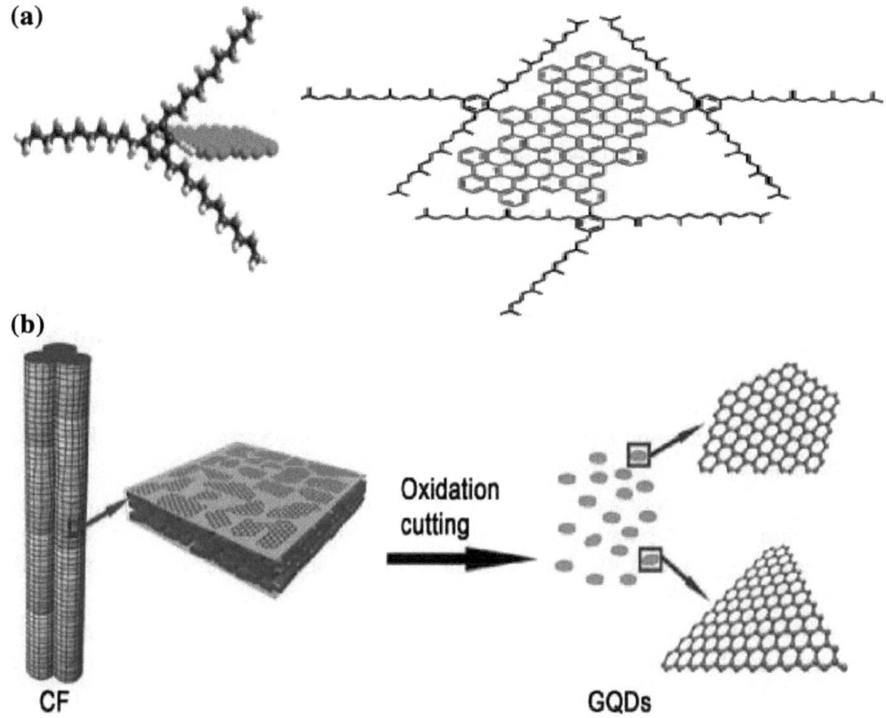

Fig. 3.2 **a** Colloidal graphene quantum dots with well-defined structure. Reprinted with permission from [25]. Copyright 2013 American Chemical Society. **b** Quantum dots obtained from graphitic fibers by oxidation cutting. Reprinted with permission from [26]. Copyright 2012 American Chemical Society

Graphene nanoribbons are 2D systems confined in one direction while quantum dots are 2D systems confined in two directions. Chemistry provides a natural route towards graphene quantum dots with up to several hundred atoms. For example, Müllen et al. used bottom-up approach from molecular nanographenes to unconventional carbon materials and a synthetic route towards easily processable and chemically tailored nanographenes on the surface of metals [9, 10, 23, 24]. Li et al. developed a chemical route toward colloidal graphene quantum dots with up to 200 carbon atoms and with well-defined structure [25], as shown in Fig. 3.2a. Ajayan et al. [26] started from graphitic fibers and used oxidation cutting to fabricate graphene quantum dots with variety of shapes, as shown in Fig. 3.2b. Berry et al. developed nanotomy-based production of transferable and dispersible graphene nanostructures of controlled shape and size [27]. Such techniques are needed if graphene quantum dots are to be used for energy-based applications, as reviewed recently by Zhang et al. [28].

For electronic and optoelectronic applications one may need quantum dots with both sizes exceeding those produced using bottom-up approaches and with full control over shape and edge type. Here, top-down techniques, including AFM, might be useful. One of the first attempts at top-down fabrication of graphene quantum dots was by McEuen et al., who studied graphite quantum dots with thickness from a few to tens of nanometers and lateral dimensions $\sim 1\ \mu m$ [29]. They were placed onto a Si wafer with a 200 nm of thermally grown oxide and connected to metallic electrodes. Transport measurements showed Coulomb blockade phenomena. By analyzing the period of Coulomb oscillations in gate voltage, they demonstrated that the dot area extends into the graphite piece lying under the electrodes. Graphene quantum dots were experimentally fabricated starting from a graphene sheet. Ponomarenko et al. produced structures with different sizes with oxygen plasma etching and a protecting mask obtained by using high-resolution electron-beam lithography [30]. Their method allowed to create quantum dots even with 10 nm radius but not with a well-defined shape. Ensslin et al. studied tunable graphene quantum dots fabricated based on reactive ion etching (RIE) patterned graphene [31–35] as shown in Fig. 3.3a. Yacoby et al. fabricated quantum dots using bilayer graphene, with the device shown in Fig. 3.3b [36]. According to an earlier prediction by Peeters et al. [37] and earlier section on bilayer graphene, application of inhomogeneous gates on top of bilayer graphene opens gaps and allows for confinement of charged carriers, as schematically indicated in Fig. 3.3b.

An alternative to previously mentioned fabrication methods is creating graphene nanostructures by cutting graphene into desired shapes. It was shown that few-layer [38] and single-layer [39] graphene can be cut by using metallic particles. The process was based on anisotropic etching by thermally activated nickel particles. The cuts were directed along proper crystallographic orientations with the width of cuts determined by a diameter of metal particles. By using this technique, they were able to produce ribbons, equilateral triangles and other graphene nanostructures.

Another method involves fabrication of graphene nanostructures using AFM [40] and direct growth on metallic surfaces. An example of a triangular graphene quantum dot grown on Ni surface is shown in Fig. 3.4a [41], graphene quantum dot on the surface of Ir in Fig. 3.4b [42] and graphene quantum dots on Cu surface in Fig. 3.4c [43].

3.2 The Role of Edges

As shown in Fig. 3.1, one can terminate the honeycomb lattice with two distinct edges: armchair and zigzag. They were experimentally observed near single-step edges on the surface of exfoliated graphite by scanning tunneling microscopy (STM) and spectroscopy (STS) [44–48] and Raman spectroscopy [49–51]. Jia et al. have shown that zigzag and armchair edges are characterized by different activation energy [4]. Their molecular dynamics calculations estimated activation energies of 11 eV for zigzag and 6.7 eV for armchair edges. This enabled them to eliminate an armchair edge in favour of zigzag edge by heating the sample with electrical current. The dynamics of edges was also studied [52, 53]. The measurements were performed in real time by side spherical aberration-corrected transmission electron microscopy

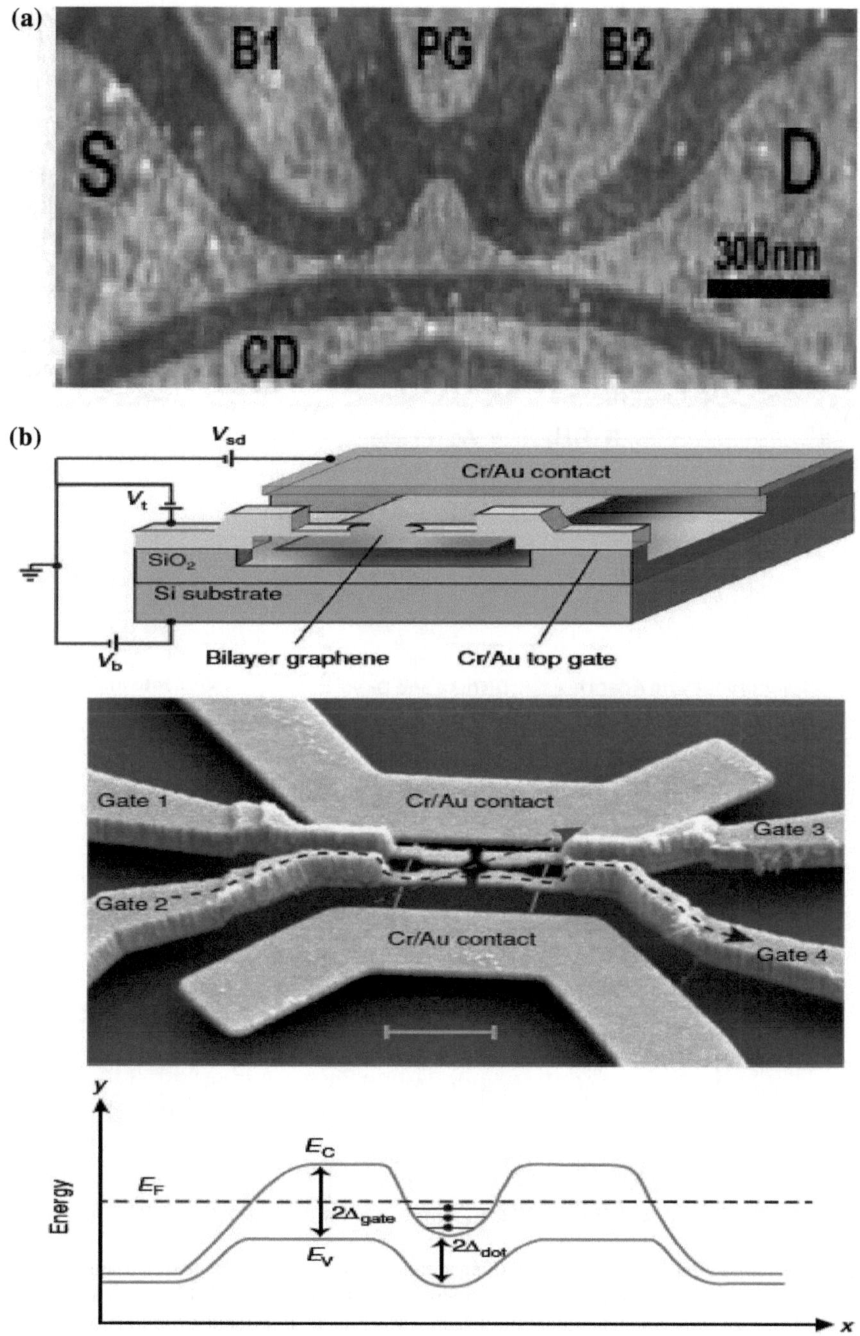

Fig. 3.3 SEM picture of **a** a quantum dot etched out of graphene, and **b** a quantum dot defined by gates in a bilayer graphene. **a** Reprinted with permission from [32]. Copyright 2008, AIP Publishing LLC. and **b** reprinted from [36]

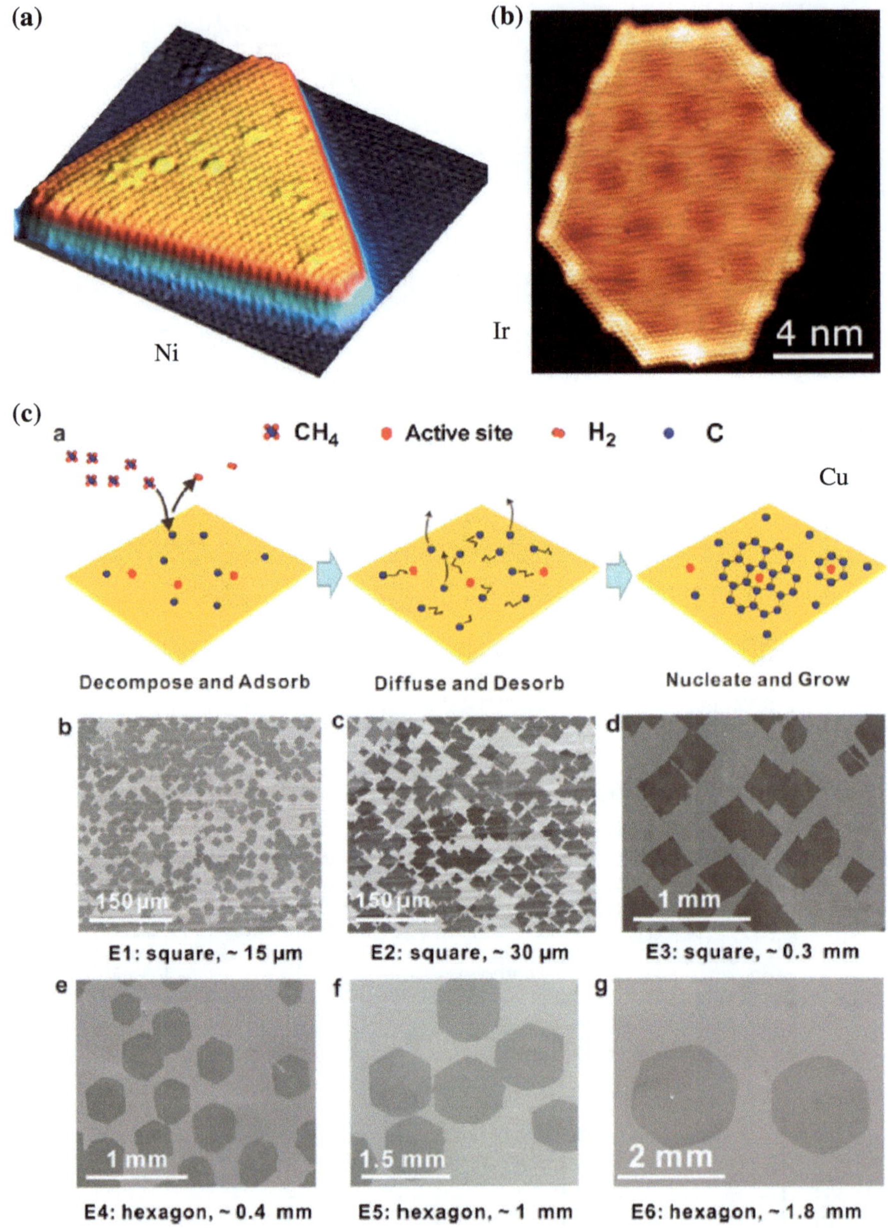

Fig. 3.4 **a** Three-dimensional rendering of an atomic resolution STM image of a triangular island of graphene on Ni(111). Reprinted with permission from [41]. Copyright 2012 American Chemical Society. **b** Image of a graphene quantum dot on surface of Ir. Reprinted from [42]. **c** Graphene quantum dots on Cu surface. Reprinted with permission from [43]. Copyright 2012 American Chemical Society

with sensitivity required to detect every carbon atom which remained stable for a sufficient amount of time. The most prominent edge structure was of the zigzag type. Koskinen, Malola and Häkkinen predicted, based on DFT calculations, the stability of reconstructed ZZ_{57} edges [54]. The variety of stable combinations of pentagons, heptagons or higher polygons was observed [53, 55].

Theoretical calculations predicted edge states in the vicinity of the Fermi energy for structures with zigzag edges [16, 56–68]. These edge states were clearly identified experimentally [44–48]. They form a degenerate band and a peak in the density of states in graphene ribbons [16, 56–58, 60]. It was also shown by using the Hubbard model in a mean-field approximation that in graphene nanoribbons the electrons occupying edge states exhibit ferromagnetic order within an edge and antiferromagnetic order between opposite zigzag edges [57, 69, 70]. Son et al. have shown by using first-principles calculations that magnetic properties can be controlled by the external electric field applied across the ribbon [58]. The electric field lifts the spin degeneracy by reducing the band gap for one spin channel and widening the gap for the other. Hence, one can change the antiferromagnetic coupling between opposite edges into the ferromagnetic one. Graphene ribbons continue to be widely investigated [71–77].

The effect of edges was also studied in graphene quantum dots (GQD). It was shown that the type of edges influences the optical properties [59, 78, 79]. In GQDs with zigzag edges, edge states can collapse to a degenerate shell on the Fermi level [59, 61–64, 66–68]. The relation between the degeneracy of the shell and the difference between the number of atoms corresponding to two graphene sublattices was pointed out [61, 62, 64, 68]. One of the systems with the degenerate shell is a triangular graphene quantum dot (TGQD). Hence, the electronic properties of TGQDs were extensively studied [12, 59, 61–64, 67, 68, 80–90]. For a half-filled degenerate shell, TGQDs were studied by Ezawa using the Heisenberg Hamiltonian [61], by Fernandez-Rossier and Palacios [62] using the mean-field Hubbard model, by Wang, Meng and Kaxiras [64] using DFT. It was shown that the ground state corresponds to fully spin-polarized edges, with a finite magnetic moment proportional to the shell degeneracy. In Chap. 5, we will investigate the magnetic properties in detail using exact diagonalization techniques [67, 90].

3.3 Size Quantization Effects

Spatial confinement of carriers in graphene nanostructures is expected to lead to the discretization of the energy spectrum and an opening of the energy gap. In graphene ribbons, the gap opening was predicted based on the tight-binding model or starting from THE Dirac Hamiltonian [56, 91, 92]. Ribbons with armchair edges oscillate between insulating and metallic ground state as the width changes. The size of the bandgap was predicted to be inversely proportional to the nanoribbon width [16]. The experimental observation indicates the opening of the energy gap for the narrowest ribbons, with scaling behavior in agreement with theoretical predictions [2, 3, 7].

Ponomarenko et al. have shown that for GQDs with a diameter $D < 100\,\text{nm}$, quantum confinement effects start playing a role [30]. They observed Coulomb blockade peak oscillations as a function of gate voltage with randomly varied peak spacings. These results were in agreement with the predictions for chaotic Dirac billiards, the expected behavior for Dirac Fermions in confinement with an arbitrary shape [93]. An exponential decrease of the energy gap as a function of the diameter for Dirac Fermions was predicted theoretically by Recher and Trauzettel [94].

In few-nm GDQs with well-defined edges, high symmetry standing waves were observed by using STM [42, 95, 96]. These observations are in good agreement with TB and DFT calculations. Akola et al. have shown that a structure of shells and super-shells in the energy spectrum of circular quantum dots and TGQD is created [63, 65]. According to their calculations, TGQD with the edge length at least $\sim$40 nm is needed to observe clearly the first super-shell. TB calculations predict an opening of the energy gap for arbitrary shape GQDs. An exponential decrease of the energy gap with the number of atoms is predicted [78, 79, 96]. This behavior is quantitatively different for structures with zigzag and armchair edges, which is related to the edge states present in systems with zigzag edges [79]. The theory of graphene quantum dots and their properties will be developed in subsequent chapters.

References

1. P. Avouris, F. Xia, MRS Bull. **37**, 1225 (2012)
2. M.Y. Han, B. Özyilmaz, Y. Zhang, P. Kim, Phys. Rev. Lett. **98**, 206805 (2007)
3. Z. Chen, Y.-M. Lin, M.J. Rooks, P. Avouris, Physica E **40**, 228 (2007)
4. X. Jia, M. Hofmann, V. Meunier, B.G. Sumpter, J. Campos-Delgado, J.M. Romo-Herrera, H. Son, Y.-P. Hsieh, A. Reina, J. Kong, M. Terrones, M.S. Dresselhaus, Science **323**, 1701 (2009)
5. M. Engelund, J.A. Fürst, A.P. Jauho, M. Brandbyge, Phys. Rev. Lett. **104**, 036807 (2010)
6. E. Cruz-Silva, A.R. Botello-Mendez, Z.M. Barnett, X. Jia, M.S. Dresselhaus, H. Terrones, M. Terrones, B.G. Sumpter, V. Meunier, Phys. Rev. Lett. **105**, 045501 (2010)
7. X. Li, X. Wang, L. Zhang, S. Lee, H. Dai, Science **319**, 1229 (2008)
8. J. Cai, P. Ruffieux, R. Jaafar, M. Bieri, T. Braun, S. Blankenburg, M. Muoth, A.P. Seitsonen, M. Saleh, X. Feng, K. Mullen, R. Fasel, Nature **466**, 470 (2010)
9. L. Zhi, K. Müllen, J. Mater. Chem. **18**, 1472 (2008)
10. M. Treier, C.A. Pignedoli, T. Laino, R. Rieger, K. Müllen, D. Passerone, R. Fasel, Nat. Chem. **3**, 61 (2010)
11. M.L. Mueller, X. Yan, J.A. McGuire, L. Li, Nano Lett. **10**, 2679 (2010)
12. Y. Morita, S. Suzuki, K. Sato, T. Takui, Nat. Chem. **3**, 197 (2011)
13. J. Lu, P.S.E. Yeo, C.K. Gan, P. Wu, K.P. Loh, Nat. Nanotechnol. **6**, 247 (2011)
14. D.V. Kosynkin, A.L. Higginbotham, A. Sinitskii, J.R. Lomeda, A. Dimiev, B.K. Prince, J.M. Tour, Nature (London) **458**, 872 (2009)
15. C. Tao, L. Jiao, O.V. Yazyev, Y.-C. Chen, J. Feng, X. Zhang, R.B. Capaz, J.M. Tour, A. Zettl, S.G. Louie, H. Dai, M.F. Crommie, Nat. Phys. **7**, 616 (2011)
16. Y.W. Son, M.L. Cohen, S.G. Louie, Phys. Rev. Lett. **97**, 216803 (2006)
17. L. Pisani, J.A. Chan, B. Montanari, N.M. Harrison, Phys. Rev. B **75**, 064418 (2007)
18. O.V. Yazyev, R.B. Capaz, S.G. Louie, Phys. Rev. B **84**, 115406 (2011)
19. O.V. Yazyev, R.B. Capaz, S.G. Louie, J. Phys. Conf. Ser. **302**, 012016 (2011)
20. H. Santos, A. Ayuela, W. Jaskólski, M. Pelc, L. Chico, Phys. Rev. B **80**, 035436 (2009)

21. L. Chico, H. Santos, A. Ayuela, W. Jaskólski, M. Pelc, L. Brey, Acta Phys. Pol. A. **118**, 433 (2010)
22. W. Jaskólski, A. Ayuela, M. Pelc, H. Santos, L. Chico, Phys. Rev. B **83**, 235424 (2011)
23. M. Müller, C. Kbel, K. Müllen, Chem. Eur. J. **4**, 2099 (1998)
24. N. Tyutyulkov, G. Madjarova, F. Dietz, K. Müllen, J. Phys. Chem. B **102**, 10183 (1998)
25. X. Yan, B. Li, L.S. Li, Acc. Chem. Res. **46**, 22542262 (2013)
26. J. Peng, W. Gao, B.K. Gupta, Z. Liu, R. Romero-Aburto, L. Ge, L. Song, L.B. Alemany, X. Zhan, G. Gao, S.A. Vithayathil, B.A. Kaipparettu, A.A. Marti, T. Hayashi, J. Zhu, P.M. Ajayan, Nano Lett. **12**, 844–849 (2012)
27. N. Mohanty, D. Moore, Z.P. Xu, T.S. Sreeprasad, A. Nagaraja, A.A. Rodriguez, V. Berry, Nat. Commun. **3**, 844 (2012)
28. Z. Zhang, J. Zhang, N. Chen, L. Qu, Energy Environ. Sci. **5**, 8869 (2012)
29. J.S. Bunch, Y. Yaish, M. Brink, K. Bolotin, P.L. McEuen, Nano Lett. **5**, 287 (2005)
30. L.A. Ponomarenko, F. Schedin, M.I. Katsnelson, R. Yang, E.W. Hill, K.S. Novoselov, A.K. Geim, Science **320**, 356 (2008)
31. C. Stampfer, J. Güttinger, F. Molitor, D. Graf, T. Ihn, K. Ensslin, Appl. Phys. Lett. **92**, 012102 (2008)
32. J. Güttinger, C. Stampfer, S. Hellmöller, F. Molitor, T. Ihn, K. Ensslin, Appl. Phys. Lett. **93**, 212102 (2008)
33. S. Schnez, J. Güttinger, M. Huefner, C. Stampfer, K. Ensslin, T. Ihn, Phys. Rev. B **82**, 165445 (2010)
34. T. Ihn, J. Güttinger, F. Molitor, S. Schnez, E. Schurtenberger, A. Jacobsen, S. Hellmüller, T. Frey, S. Dröscher, C. Stampfer, K. Ensslin, Mater. Today **13**, 44 (2010)
35. F. Molitor, J. Güttinger, C. Stampfer, S. Dröscher, A. Jacobsen, T. Ihn, K. Ensslin, J. Phys. Condens. Matter **23**, 243201 (2011)
36. M.T. Allen, J. Martin, A. Yacoby, Nat. Commun. **3**, 934 (2012)
37. J.M. Pereira Jr, P. Vasilopoulos, F.M. Peeters, Nano Lett. **7**, 946 (2007)
38. L. Ci, Z. Xu, L. Wang, W. Gao, F. Ding, K.F. Kelly, B.I. Yakobson, P.M. Ajayan, Nano Res. **1**, 116 (2008)
39. L.C. Campos, V.R. Manfrinato, J.D. Sanchez-Yamagishi, J. Kong, P. Jarillo-Herrero, Nano Lett. **9**, 2600 (2009)
40. L. Tapazsto, G. Dobrik, P. Lambin, L.P. Bir, Nat. Nanotechnol. **3**, 397 (2008)
41. M. Olle, G. Ceballos, D. Serrate, P. Gambardella, Nano Lett. **12**, 4431 (2012)
42. S.K. Hämäläinen, Z. Sun, M.P. Boneschanscher, A. Uppstu, M. Ijäs, A. Harju, D. Vanmaekelbergh, P. Liljeroth, Phys. Rev. Lett. **107**, 236803 (2011)
43. Z. Yan, J. Lin, Z. Peng, Z. Sun, Y. Zhu, L. Li, C. Xiang, E.L. Samuel, C. Kittrell, J.M. Tour, ACS Nano **6**, 9110 (2012)
44. Y. Niimi, T. Matsui, H. Kambara, K. Tagami, M. Tsukada, H. Fukuyama, Appl. Surf. Sci. **241**, 43 (2005)
45. Y. Kobayashi, K.-I. Fukui, T. Enoki, K. Kusakabe, Y. Kaburagi, Phys. Rev. B **71**, 193406 (2005)
46. Y. Kobayashi, K. Fukui, T. Enoki, K. Kusakabe, Phys. Rev. B **73**, 125415 (2006)
47. Y. Niimi, T. Matsui, H. Kambara, K. Tagami, M. Tsukada, H. Fukuyama, Phys. Rev. B **73**, 085421 (2006)
48. C. Tao, L. Jiao, O.V. Yazyev, Y.-C. Chen, J. Feng, X. Zhang, R.B. Capaz, J.M. Tour, A. Zettl, S.G. Louie, H. Dai, M.F. Crommie, Nature Phys. **7**, 616 (2011)
49. Y. You, Z. Ni, T. Yu, Z. Shen, Appl. Phys. Lett. **93**, 163112 (2008)
50. B. Krauss, P. Nemes-Incze, V. Skakalova, L.P. Biró, K. von Klitzing, J.H. Smet, Nano Lett. **10**, 4544 (2010)
51. S. Neubeck, Y.M. You, Z.H. Ni, P. Blake, Z.X. Shen, A.K. Geim, K.S. Novoselov, Appl. Phys. Lett. **97**, 053110 (2010)
52. Ç.Ö. Girit, J.C. Meyer, R. Erni, M.D. Rossell, C. Kisielowski, L. Yang, C. Park, M.F. Crommie, M.L. Cohen, S.G. Louie, A. Zettl, Science **323**, 1705 (2009)
53. A. Chuvilin, J.C. Meyer, G. Algara-Siller, U. Kaiser, New J. Phys. **11**, 083019 (2009)
54. P. Koskinen, S. Malola, H. Häkkinen, Phys. Rev. Lett. **101**, 115502 (2008)

55. P. Koskinen, S. Malola, H. Häkkinen, Phys. Rev. B **80**, 073401 (2009)
56. K. Nakada, M. Fujita, G. Dresselhaus, M.S. Dresselhaus, Phys. Rev. B **54**, 17954 (1996)
57. M. Fujita, K. Wakabayashi, K. Nakada, K. Kusakabe, J. Phys. Soc. Jpn. **65**, 1920 (1996)
58. Y. Son, M.L. Cohen, S.G. Louie, Nature **444**, 347 (2006)
59. T. Yamamoto, T. Noguchi, K. Watanabe, Phys. Rev. B **74**, 121409 (2006)
60. M. Ezawa, Phys. Rev. B **73**, 045432 (2006)
61. M. Ezawa, Phys. Rev. B **76**, 245415 (2007)
62. J. Fernandez-Rossier, J.J. Palacios, Phys. Rev. Lett. **99**, 177204 (2007)
63. J. Akola, H.P. Heiskanen, M. Manninen, Phys. Rev. B **77**, 193410 (2008)
64. W.L. Wang, S. Meng, E. Kaxiras, Nano Lett. **8**, 241 (2008)
65. M. Manninen, H.P. Heiskanen, J. Akola, Eur. Phys. J. D **52**, 143146 (2009)
66. W.L. Wang, O.V. Yazyev, S. Meng, E. Kaxiras, Phys. Rev. Lett. **102**, 157201 (2009)
67. A.D. Güçlü, P. Potasz, O. Voznyy, M. Korkusinski, P. Hawrylak, Phys. Rev. Lett. **103**, 246805 (2009)
68. P. Potasz, A.D. Güçlü, P. Hawrylak, Phys. Rev. B **81**, 033403 (2010)
69. K. Wakabayashi, M. Sigrist, M. Fujita, J. Phys. Soc. Jpn. **67**, 2089 (1998)
70. A. Yamashiro, Y. Shimoi, K. Harigaya, K. Wakabayashi, Phys. Rev. B **68**, 193410 (2003)
71. L. Yang, M.L. Cohen, S.G. Louie, Phys. Rev. B **101**, 186401 (2008)
72. F. Cervantes-Sodi, G. Csanyi, S. Piscanec, A.C. Ferrari, Phys. Rev. B **77**, 165427 (2008)
73. M. Fernǻndez-Rossier, Phys. Rev. B **77**, 075430 (2008)
74. J.J. Palacios, J. Fernǻndez-Rossier, L. Brey, Phys. Rev. B **77**, 195428 (2008)
75. J. Jung, A.H. MacDonald, Phys. Rev. B **79**, 235433 (2009)
76. D. Soriano, J. Fernǻndez-Rossier, Phys. Rev. B **82**, 161302(R) (2010)
77. H. Wang, V.W. Scarola, Phys. Rev. B **85**, 075438 (2012)
78. Z.Z. Zhang, K. Chang, F.M. Peeters, Phys. Rev. B **77**, 235411 (2008)
79. A.D. Güçlü, P. Potasz, P. Hawrylak, Phys. Rev. B **82**, 155445 (2010)
80. M. Ezawa, Phys. Rev. B **77**, 155411 (2008)
81. M.R. Philpott, F. Cimpoesu, Y. Kawazoe, Chem. Phys. **354**, 1 (2008)
82. H.P. Heiskanen, M. Manninen, J. Akola, New J. Phys. **10**, 103015 (2008)
83. D.P. Kosimov, A.A. Dzhurakhalov, F.M. Peeters, Phys. Rev. B **81**, 195414 (2008)
84. M. Ezawa, Phys. Rev. B **81**, 201402 (2010)
85. M. Ezawa, Phys. E **42**, 703 (2010)
86. H. Sahin, R.T. Senger, S. Ciraci, J. Appl. Phys. **108**, 074301 (2010)
87. I. Romanovsky, C. Yannouleas, U. Landman, Phys. Rev. B **83**, 045421 (2011)
88. M. Zarenia, A. Chaves, G.A. Farias, F.M. Peeters, Phys. Rev. B **84**, 245403 (2011)
89. O. Voznyy, A.D. Güçlü, P. Potasz, P. Hawrylak, Phys. Rev. B **83**, 165417 (2011)
90. P. Potasz, A.D. Güçlü, A. Wǿjs, P. Hawrylak, Phys. Rev. B **85**, 075431 (2012)
91. L. Brey, H.A. Fertig, Phys. Rev. B **73**, 195408 (2006)
92. L. Brey, H.A. Fertig, Phys. Rev. B **73**, 235411 (2006)
93. M.V. Berry, R.J. Mondragon, Proc. R. Soc. Lond. A **412**, 53–74 (1987)
94. P. Recher, B. Trauzettel, Nanotechnology **21**, 302001 (2010)
95. D. Subramaniam, F. Libisch, Y. Li, C. Pauly, V. Geringer, R. Reiter, T. Mashoff, M. Liebmann, J. Burgdörfer, C. Busse, T. Michely, R. Mazzarello, M. Pratzer, M. Morgenstern, Phys. Rev. Lett. **108**, 046801 (2012)
96. K.A. Ritter, J.W. Lyding, Nat Mater. **8**, 235 (2009)

Chapter 4
Single-Particle Properties of Graphene Quantum Dots

Abstract This chapter describes the size, shape and edge dependence of the electronic properties of graphene quantum dots obtained using the empirical tight-binding model. The effective mass extension of the TB model is discussed, including the effect of the magnetic field. The one-band TB model is extended to the sp^2 TB model and spin-orbit coupling is introduced, followed by the Kane-Mele Hamiltonian and the spin Hall effect in nanoribbons. Triangular quantum dots and rings with zigzag edges as examples of quantum dots with broken sublattice symmetry and a shell of degenerate states at the Fermi level are described. Graphene ribbons and twisted graphene Möbius ribbons as examples of topological insulators where topology is introduced through geometry are discussed.

4.1 Size, Shape and Edge Dependence of Single Particle Spectrum

We discuss here how single particle properties of graphene can be engineered by varying the size, shape, type of edge, sublattice symmetry and number of layers. In the following chapters the important effect of electron-electron interactions will be discussed.

4.1.1 One-Band Empirical Tight-Binding Model

The one-band empirical tight-binding model introduced by Wallace [1] and discussed in Sect. 2.4 describes successfully the one-electron spectrum of bulk graphene. Here we discuss predictions of the tight-binding model applied to graphene quantum dots (GQDs). Within the one-band p_z model we do not consider explicitly the sp^2 hybridized orbitals at the edges, we assume passivation of edges with hydrogen and defer the discussion of edge passivation and stability to the section on the four-band model in this chapter and on ab-initio theory in the following chapter.

We start by expanding the wavefunction of electron in terms of p_z orbitals localized on carbon atoms. Either neglecting overlap of p_z orbitals on neighboring atoms

A.D. Güçlü et al., *Graphene Quantum Dots*,
NanoScience and Technology, DOI 10.1007/978-3-662-44611-9_4

or starting from Wannier orthogonal orbitals, the simplest tight-binding Hamiltonian in the second quantization form with only nearest neighbor hopping included can be written as

$$H_{TB} = t \sum_{\langle i,l \rangle,\sigma} c_{i\sigma}^{\dagger} c_{l\sigma}, \tag{4.1}$$

where $c_{i\sigma}^{\dagger}$ and $c_{i\sigma}$ are creation and annihilation operators for an electron on the lattice site i with spin σ, and $\langle i, l \rangle$ indicates a summation over nearest neighbor sites. The negative hopping integral t between nearest A and B neighbor atoms corresponding to two sublattices is defined in (2.14). The TB Hamiltonian can describe finite-size systems by restricting the tunneling matrix elements to atoms within the quantum dot. We describe a method of building the TB Hamiltonian matrix on the example of a hexagonal quantum dot, shown in Fig. 4.1b, consisting of $N_A = 12$ A and $N_B = 12$ B atoms, with a total of $N = 24$ atoms. The quantum dot is constructed starting with a benzene ring (Fig. 4.1a), and adding one more ring of benzene molecules. The edge of this quantum dot is a zigzag edge consisting of equal number of A and B atoms having only two nearest carbon neighbors instead of three as in bulk graphene. We note that the next in size hexagonal dot (Fig. 4.1c), has armchair edge and $N = 42$ atoms. The positions of all sublattice A and B atoms in the quantum dot are given by $\mathbf{R}_A = n\mathbf{a}_1 + m\mathbf{a}_2$ and $\mathbf{R}_B = n\mathbf{a}_1 + m\mathbf{a}_2 + \mathbf{b}$, where n and m are integers, $\mathbf{b}$ is a vector going from the A atom to the B atom in a unit cell, and $\mathbf{R}_A$ and $\mathbf{R}_B$ are within the predefined area of the quantum dot.

The two indices (n, m) describing the position of each atom are translated into atom indices j, from $j = 1$ to $j = 24$, as shown in Fig. 4.1b. The electron wavefunction is a linear combination of the 24 p_z orbitals localized on carbon atoms and the TB Hamiltonian is a 24×24 matrix. The nonzero matrix elements of the TB Hamiltonian given by (4.1) correspond to tunneling matrix element t between orbitals on neighboring sites. According to Fig. 4.1, the carbon atom 1 is connected to atoms 2, 6 and 7. Hence, the first row of the Hamiltonian matrix, which describes tunneling out of carbon atom 1, contains all zeros except for 2nd, 6th and 7th columns where we have t. All remaining rows can be constructed in a similar manner. Once the Hamiltonian matrix is built, it is diagonalized numerically yielding eigenvalues and eigenvectors labeled by indices from "1" to "24". In Fig. 4.2, we show the energy spectra, eigenvalues $E(i)$, as a function of eigenstate index $i = 1, 2, \ldots, 24$ obtained by diagonalization of the Hamiltonian matrix. Without tunneling all energy levels were degenerate, $E(i) = 0$. We see that tunneling removed the degeneracy and led to the formation of the band of 12 valence states below the Fermi level, band of 12 conduction states above the Fermi level, and a gap E_g across the Fermi level. The energy spectrum of the graphene quantum dot is now similar to the energy spectrum of semiconductor nanocrystals and quantum dots [2–4]. However, we see that the top of the valence band and the bottom of the conduction band consists of two degenerate levels. This is to be contrasted with gated and self-assembled 2D semiconductor quantum dots, where electron and hole energy spectra form electronic shells of 2D harmonic oscillator [5], starting with a nondegenerate s-shell, followed by a doubly

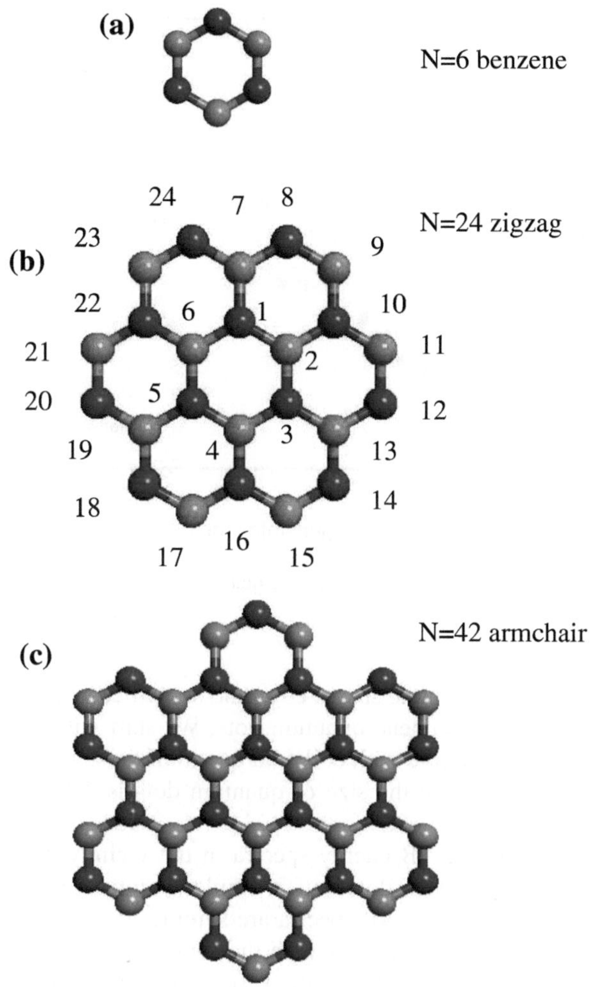

Fig. 4.1 Example of a hexagonal graphene quantum dot with $N = 24$ atoms and zigzag edge (**b**), starting with a benzene ring (**a**) and ending with larger quantum dot with $N = 42$ atoms and armchair edge (**c**)

degenerate p-shell etc. In the effective mass model discussed in the next section the double degeneracy is traced to the two non-equivalent K points, and hence two types of Dirac Fermions characterized by the valley index. The double degeneracy of the HOMO levels starts with the benzene ring where the two degenerate levels correspond to the electron moving either to the left or to the right. The degeneracy of the LUMO level follows from electron-hole symmetry. Similar arguments follow for the $N = 24$ quantum dot which consists of the benzene $N = 6$ atom ring and the $N = 18$ atom outer ring. The HOMO level corresponds to the electron moving to the left or right on the edge and central rings. In this way it appears that the valley degeneracy persists down to very small quantum dots.

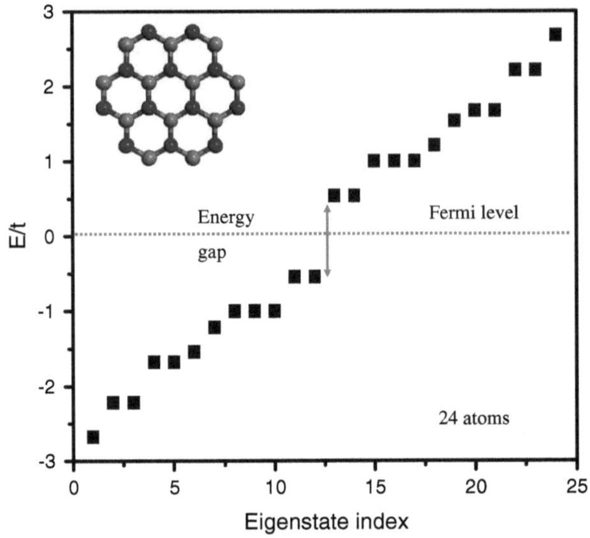

Fig. 4.2 Energy spectrum of a hexagonal graphene quantum dot with $N = 24$ atoms and zigzag edge

We now turn to discussing the shape, edge and sublattice symmetry dependence of the energy spectra of graphene quantum dots. We start with sizes of the order of $N \approx 100$ atoms, compatible with colloidal quantum dots. The discussion of the dependence of energy gap on the size of quantum dots is deferred to Chap. 7 on optical properties.

In Fig. 4.3, we show the TB energy spectra in the vicinity of the Fermi level, $E = 0$, for graphene quantum dots with a similar number of atoms, $N \sim 100$, but different shapes and edges. All spectra are symmetric with respect to $E = 0$. Figure 4.3a, b correspond to structures with the same armchair edge but different, hexagonal or triangular, shape. Both quantum dots contain the same number of A and B atoms and the sublattice symmetry is preserved. As a result of size quantization, an energy gap opens with a comparable magnitude in both quantum dots. The energy spectra look almost identical, with very similar shells of energy levels. Starting from the Fermi level, we again observe first a doubly-degenerate state, next two single and two degenerate levels in both cases. Thus, one can conclude that the shape of small graphene quantum dots with armchair edges does not play an important role for the energy spectrum in the vicinity of the Fermi level. The differences appear for larger structures and will be discussed in the section on effective mass.

In Fig. 4.3c, d, we show the energy spectra for structures with identical zigzag edge but with different shape, hexagonal and triangular. Hence, the Fig. 4.3 allows us to compare both the edge dependence for the same shape of a quantum dot and the shape dependence for the same edge. Comparing the energy gaps in hexagonal dots, the energy gap for a quantum dot with the zigzag edge is smaller compared to the energy gap of the hexagonal armchair quantum dots. Comparing the energy spectra

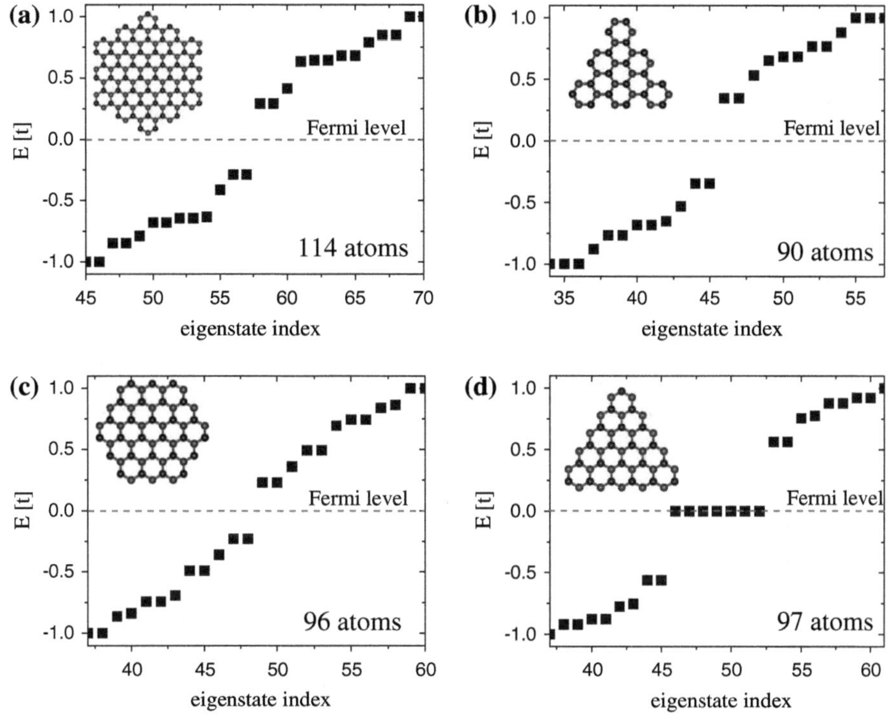

Fig. 4.3 TB energy spectra in the vicinity of the Fermi level, $E = 0$, for graphene quantum dots with a similar number of atoms, $N \sim 100$, but different shapes and edges. Energy spectra for **a** hexagonal and **b** triangular quantum dots with armchair edges, and for **c** hexagonal and **d** triangular quantum dots with zigzag edges. Edge effects appear only in systems with zigzag edges

of hexagonal and triangular quantum dot (TGQD) with zigzag edges one finds that the deformation of a hexagon to a triangle led to a dramatic rearrangement of the energy spectrum. The energy spectrum in the vicinity of the Fermi level collapsed to a degenerate shell at the Fermi level. The degenerate shell is related to the broken sublattice symmetry—changing shape from a hexagon to a triangle requires removing a number of carbon atoms and breaks sublattice symmetry. A detailed analysis of the energy spectra of TGQDs will be presented in Sect. 4.3.

In Fig. 4.4, we show the probability densities of the highest valence energy levels corresponding to quantum dots with the energy spectra shown in Fig. 4.3. In all four quantum dots these states are doubly degenerate, thus we plot the sum of probability densities for these two states. The electronic probability densities defined in this way preserve the symmetry of quantum dots. We also note that identical electronic densities are obtained for the lowest energy levels from the conduction band. Eigenfunctions for a valence state Ψ_v with an energy $E_v = -|E|$ and for a conduction state Ψ_c with an energy $E_c = |E|$ are identical on lattice sites corresponding to sublattice A, and have opposite signs on lattice sites corresponding to sublattice B. The valence band represents bonding, and conduction band represents anti-bonding

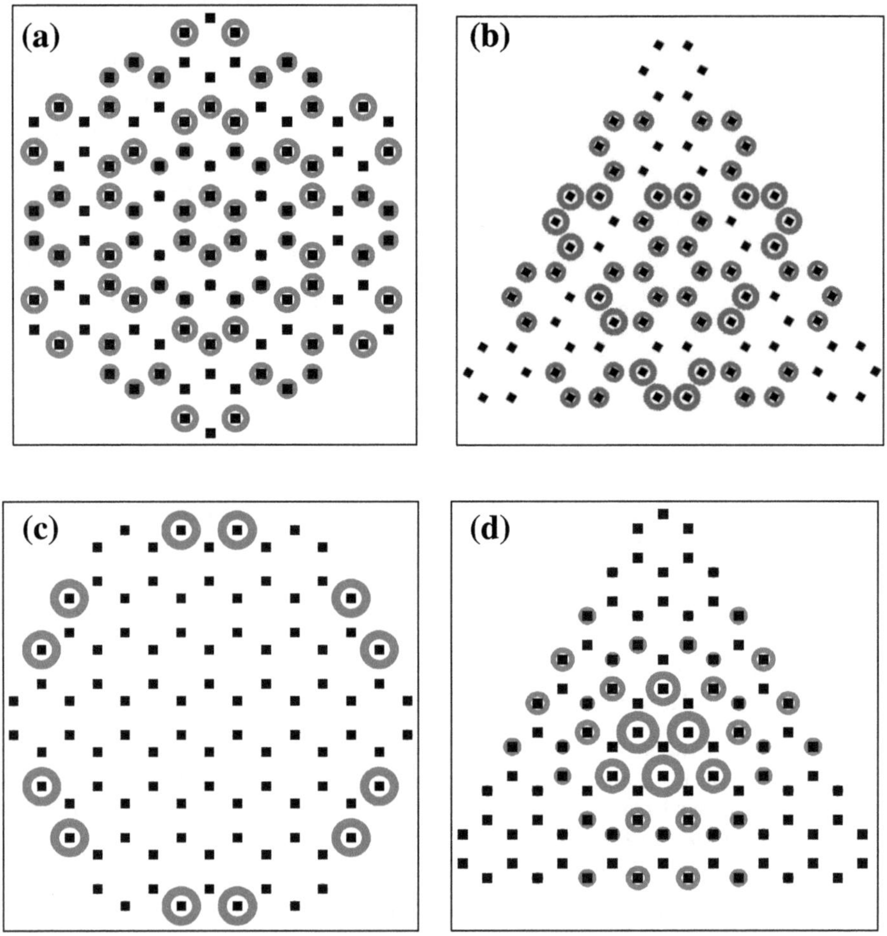

Fig. 4.4 Electronic probability densities of the highest valence energy levels corresponding to structures with the energy spectra shown in Fig. 4.3. Only in hexagonal structure with zigzag edges (**c**), these states are edge states. (**a-d**) The radius of circles is proportional to the electronic probability density on atomic sites marked by black squares

states of two sublattices. Thus, the electronic probability densities are identical in both cases. For the hexagonal structure with armchair edges (Fig. 4.4a), the electronic density spreads over the entire structure. Starting from the center, alternating hexagons with an increasing size characterized by higher and lower densities are seen. In the triangular structure with armchair edges (Fig. 4.4b), the electronic density is localized in the center of the structure, avoiding corners. A large concentration of the density with a triangular shape rotated by $\frac{\pi}{6}$ with respect to the corners is observed. In Fig. 4.4c, the electronic density of valence states for the hexagonal dot with zigzag edges is plotted. These states are strongly localized on six edges. If we are to think of the states localized at the edges as corresponding to electrons moving at the edge, quantization of their energy levels is given by the circumference of the edge and not the diameter. Hence, edge effects are responsible for faster closing of the energy gap with increasing size in comparison to quantum dots with armchair edges. This statement can be confirmed by comparing the energy gaps from Fig. 4.3c with a and b. The energy gap as a function of size will be studied in detail in Sec. 5.1. On the other hand, no edge effects are observed in Fig. 4.4d, in TGQD with zigzag edges. Here, the electronic density of the highest valence states is localized in the center of the structure. However, in this system a degenerate shell appears. In Sect. 2.3.1 we show that edge states in TGQD collapse to this degenerate shell. We note that similar patterns of electronic probability densities plotted in Fig. 4.4 were observed in larger structures for quantum dots with different shapes.

Increasing the number of atoms in a GQD to several hundred allows us to examine the density of states (DOS) $D(E)$ and compare with bulk density of states. In Fig. 4.5 the density of states $D(E) = \sum_i \delta(E(i) - E)$ for GQD consisting of $N \approx 600$ atoms with different shapes and edges are plotted. Due to a similarity between the energy spectra from Fig. 4.3a, b, for hexagonal and triangular dots with armchair edges only the DOS for the first one is shown. In order to smooth the discrete energy spectra, we use the Gaussian broadening function $f(E) = \exp(-(E - E_i)^2/\Gamma^2)$ of each energy level E_i with a width $\Gamma = 0.024|t|$. DOS for a GQD with armchair edges and $N = 546$ atoms vanishes close to the Fermi energy, $E = 0$, in analogy with the infinite graphene (not shown) [6]. GQDs with zigzag edges have an additional contribution from edge states, seen as peaks at zero energy. The peak for TGQD, $N = 622$ atoms, with a zigzag edge is significantly higher compared with the hexagonal dot with a

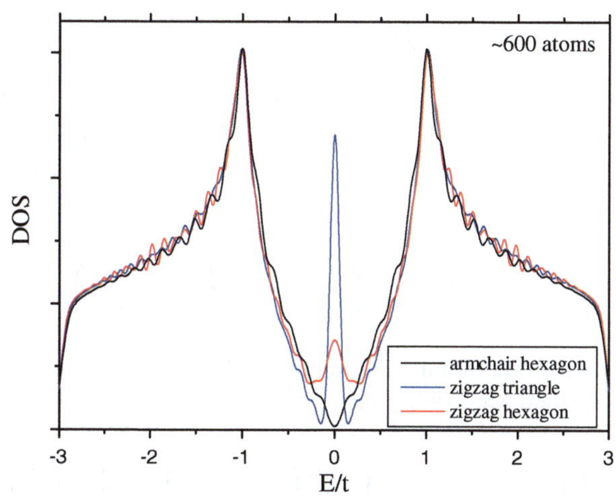

Fig. 4.5 The density of states (DOS) for GQD consisting of around $N = 600$ atoms with different shapes. DOS for the system with armchair edges vanishes close to the energy $E = 0$, in analogy with infinite graphene. Graphene quantum dots with zigzag edges have an additional contribution from edge states, seen as a peak at $E = 0$ point

zigzag edge, $N = 600$ atoms, due to a collapse of edge states to the degenerate shell at zero energy $E = 0$. Farther away from the Fermi level, the DOS looks similar for quantum dots with all shapes and is comparable to DOS for infinite graphene, with characteristic van Hove singularities at $E = \pm t$ [6].

4.1.2 Effective Mass Model of Graphene Quantum Dots

The energy spectra of Dirac Fermions in large, sub-micron scale graphene quantum dots with millions of atoms can be understood using the effective mass Hamiltonian of graphene and appropriate boundary conditions. We consider a Dirac Fermion on a circular 2D disk subject to a potential V described by the Hamiltonian [7, 8]

$$H_D = v_F \mathbf{p} \cdot \sigma + \tau V(r)\sigma_z, \tag{4.2}$$

where v_F is the Fermi velocity, $\sigma = (\sigma_x, \sigma_y)$ are Pauli's spin matrices in the basis of the two sublattices of A and B atoms. $V(r)$ is a mass potential coupled to the Hamiltonian via the σ_z Pauli matrix. We consider the case where $V(r) = 0$ for $r < R$, where R is a radius of the dot. At the edge of the dot, $V(R) \to \infty$. The parameter $\tau = \pm 1$ distinguishes the two inequivalent valleys K and K'. The Dirac equation given by (4.2) in cylindrical coordinates is written as

$$\begin{aligned} - i\hbar v_F \begin{pmatrix} 0 & e^{-i\phi}(\partial_r - i\partial_\phi/r) \\ e^{i\phi}(\partial_r + i\partial_\phi/r) & 0 \end{pmatrix} \begin{pmatrix} \Psi_A(r,\phi) \\ \Psi_B(r,\phi) \end{pmatrix} \\ = E \begin{pmatrix} \Psi_A(r,\phi) \\ \Psi_B(r,\phi) \end{pmatrix} \end{aligned} \tag{4.3}$$

where Ψ_A and Ψ_B are the sublattice components of the wavefunctions Ψ. The circular symmetry of the dot ensures the conservation of the total angular momentum J_z. Thus, $[H_D, J_z] = 0$, and eigenfunctions can be written in terms of angular and radial components as

$$\Psi(r,\phi) = \begin{pmatrix} \Psi_A(r,\phi) \\ \Psi_B(r,\phi) \end{pmatrix} = e^{im\phi} \begin{pmatrix} \chi_A(r) \\ e^{i\phi}\chi_B(r) \end{pmatrix}. \tag{4.4}$$

where $\chi(r)$ represents the radial components of Ψ. The eigenstates of Dirac Hamiltonian can be classified according to J_z,

$$J_z \Psi(r,\phi) = \hbar \left(-i\partial_\phi \times 1 + \frac{1}{2}\sigma_z \right) e^{im\phi} \begin{pmatrix} \chi_A(r) \\ e^{i\phi}\chi_B(r) \end{pmatrix} = \hbar \left(m + \frac{1}{2} \right) \Psi(r,\phi), \tag{4.5}$$

where $\chi_{A,B}(r)$ are Bessel functions which, in order to satisfy (4.3), are written as $\chi_A(r) = \chi_m(kr)$ and $\chi_B(r) = \chi_{m+1}(kr)$ with $k = E/\hbar v_F$. Since the Dirac Fermions are confined inside the quantum dot, we require a vanishing current at the edge [7]. This leads to the infinite-mass boundary condition which, for a circular confinement, gives

$$\Psi_B(kR)/\Psi_A(kR) = i\tau \exp i\phi. \tag{4.6}$$

This gives the following restriction on allowed values of k_{nm}

$$\tau \chi_m(k_{nm}R) = \chi_{m+1}(k_{nm}R). \tag{4.7}$$

We note that this is a different condition than for the Schrödinger electron for which the Bessel function has to vanish at the edge. The energy spectrum can be written in terms of quantized wavevectors k_{nm} as

$$E_{n,m} = \hbar v_F k_{nm}. \tag{4.8}$$

From the property of Bessel functions, $\chi_m(x) = (-1)^m \chi_{-m}(x)$, one gets $E_{n,m}(\tau) = E_{n,-m-1}(-\tau)$, which means that each energy level is doubly degenerate due to the presence of two valleys, K and K'. In Fig. 4.6 we show the energy spectrum of the circular quantum dot with radius $R = 25$ nm. Energies are written in units of the hopping integral t, $|t| = 2.5$ eV, from the TB model. We see that the confinement of the Dirac Fermion opened an energy gap in the energy spectrum. The energy gap, $E_g = 2\hbar v_F k_0 R$ (with $v_F = 3|t|a/2\hbar$), is given by $k_0 R$ which is a solution of $\chi_0(k_0 R) = \chi_1(k_0 R)$ and $\chi_0(x)$ are Bessel functions J ($\chi_0(x) = J_0(x)$). One finds $k_0 R = 1.435$ and $E_g = 2.87\hbar v_F/R$. The energy gap opens at the Fermi energy, with a magnitude inversely proportional to the radius of the quantum dot.

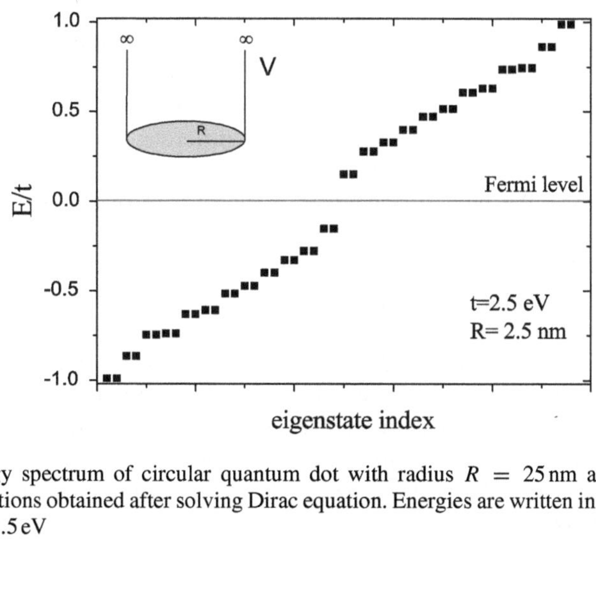

Fig. 4.6 Energy spectrum of circular quantum dot with radius $R = 25$ nm and infinite mass boundary conditions obtained after solving Dirac equation. Energies are written in units of hopping integral, $|t| = 2.5$ eV

Different boundary conditions in (4.6) were also proposed [9]. Wunsch et al. considered a GQD with a circular geometry and a zigzag graphene edge ending always on atoms belonging to the same sublattice. For a spinor function from (4.3) such boundary condition can be written as $\Psi^A(R, \theta) = 0$ [9]. The band of degenerate edge states close to the Fermi energy, similarly to the TGQD zero-energy shell, was observed. Degenerate zero-energy states also appear in a parabolic magnetic quantum dot [10]. Comparisons between the TB model and effective mass results were also performed [11–13]. Rozhkov and Nori provided an exact solution for the Dirac Fermion in a triangular quantum dot with armchair edges [14] while Wimmer et al. studied Dirac Fermions in quantum dots with disordered edges [15].

The energy levels for a Schrödinger electron in a triangular cavity are given by the equation [16, 17]

$$E_{n,m} = \varepsilon_0(n^2 + m^2 - nm), \tag{4.9}$$

where $\varepsilon_0 = 8\pi^2\hbar^2/3m_eL^2$ with m_e being the electron mass and L the length of one edge. The energy levels of a massless Dirac electron are given by [11]

$$E_{n,m} = \varepsilon_1\sqrt{n^2 + m^2 - nm}, \tag{4.10}$$

where $\varepsilon_1 = 2\pi t/\sqrt{3N}$. For the zigzag triangle, we have restrictions $m > 1$ and $n > 2m$, whereas for the armchair triangle we do have the levels where $n = m$. These special levels, called "ghost states" [11], correspond to an additional sequence of levels that do not appear for free massless particles confined in a triangular cavity. Figure 4.7 compares the density of states for the Schrödinger and Dirac electron. Appearance of the supershell structure is observed in Fig. 4.7.

Geometrical effects were further investigated by Zarenia et al. [13]. They compared the energy spectra of quantum dots with triangular and hexagonal shapes obtained within the TB model and by numerically solving the Dirac equation. For a continuum model, three types of boundary conditions were invoked: zigzag, armchair, and infinite-mass, with energy spectra as a function of the dot area shown in Fig. 4.8. The energy spectra look qualitatively different. The spectra for armchair and infinite-mass boundary conditions are significantly different from the spectra obtained for the GQDs with zigzag edges. A closer comparison with the TB model shows that the infinite-mass boundary condition may not give a good description of hexagonal GQDs, but appears satisfactory for armchair triangles. In the case of TGQD with zigzag edges (Fig. 4.8b), the existence of the zero-energy shell is captured by the continuum model with zigzag edges. A more quantitative comparison shows that the continuum model does not predict the correct number of zero-energy states, as shown in Fig. 4.9. In the continuum model the number of the degenerate states is overestimated. Similar situation takes place in the case of the number of edge states in the zigzag hexagonal graphene dot (not shown here). On the other hand, the energy gap E_g as a function of the size of the dot shown in the inset of Fig. 4.9a is comparable in both models.

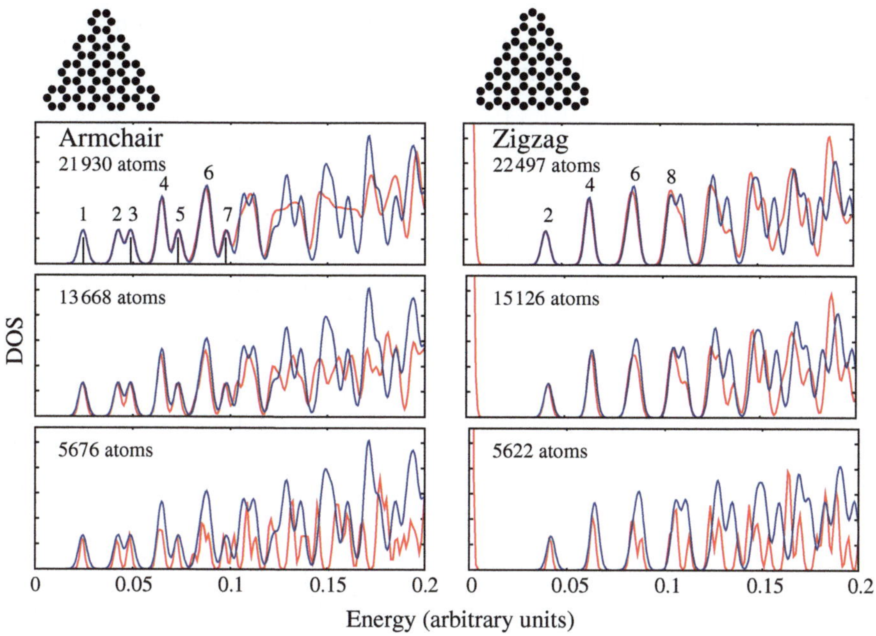

Fig. 4.7 TB-DOS above the Fermi energy for triangular flakes (*red curves*), compared to the DOS of (4.10) (*blue curves*). The sizes of the *triangles* are given as numbers of atoms. Note that the *triangles* schematically showing the geometries are much smaller from the ones used for the computation of the DOS. The energy is in units of t for the largest armchair and zigzag triangles, respectively. For the smaller sizes, the energy has been scaled by the square root of the number of atoms in order to get the peaks at the same positions. Reprinted from [11]

4.1.3 Graphene Quantum Dots in a Magnetic Field in the Effective Mass Approximation

Application of a magnetic field perpendicular to the plane of the quantum dot leads to new effects which do not take place in semiconductor quantum dots. In semiconductor quantum dots size quantization increases the band gap and introduces energy shells for both electrons and holes. In a magnetic field the energy gap increases and electronic shells convert to equally energetically spaced Landau levels in the conduction and valence band [5]. Graphene is a semimetal and application of a magnetic field leads to Landau quantization of conduction and valence band states, whose energy increases with the magnetic field. In addition, there exists a new, $n = 0$, Landau level which combines electron and hole states from the vicinity of the Fermi level [18]. The density of states of this anomalous Landau level increases with magnetic field but its energy does not change with the magnetic field and remains at the Fermi level, $E = 0$. When graphene is reduced to a quantum dot, a gap opens up. When a magnetic field is applied, some of the states of the conduction and of the valence band must evolve into the $n = 0$ Landau level and hence the gap closes with

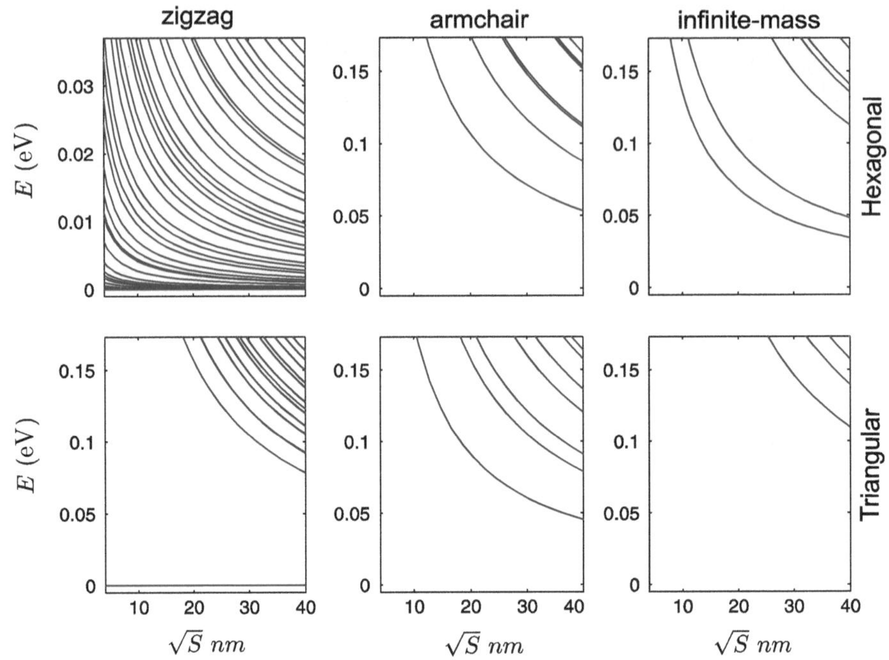

Fig. 4.8 Energy levels of hexagonal [(**a**), (**c**), and (**e**)] and triangular [(**b**), (**d**), and (**f**)] graphene quantum dots with zigzag [(**a**) and (**b**)], armchair [(**c**) and (**d**)] edges and infinite-mass boundary condition [(**e**) and (**f**)] as a function of the square root of the dot area S. Reprinted from [13]

an increasing magnetic field in contrast with semiconductor quantum dots. These qualitative considerations will now be quantitatively verified and illustrated.

The behavior of energy spectra of graphene quantum dots in an external magnetic field were studied by both the continuum effective mass Dirac Hamiltonian and TB models [8, 15, 19–25]. Schnez et al. considered a massless Dirac electron confined in a circular potential in the presence of an external magnetic field [8]. The Dirac Hamiltonian was written as

$$H_D = v_F\,(\mathbf{p} + e\mathbf{A}) \cdot \sigma + \tau V(r)\sigma_z, \tag{4.11}$$

where the vector potential $\mathbf{A} = B/2(-r\sin\phi, \cos\phi, 0)$ is written in a symmetric gauge in cylindrical coordinates (r, ϕ). Following Schnez et al., a solution of the Dirac equation with the Hamiltonian given by (4.11) will be found. Using eigenfunctions of the total angular momentum operator, (4.4), the solution of (4.11) for one of spinor components can be written as

$$\left[\partial_r^2 + \frac{1}{r}\partial_r - \frac{m+1}{l_B^2} - \frac{m^2}{r^2} - \frac{r^2}{4l_B^4} + k^2\right]\chi_A(r) = 0, \tag{4.12}$$

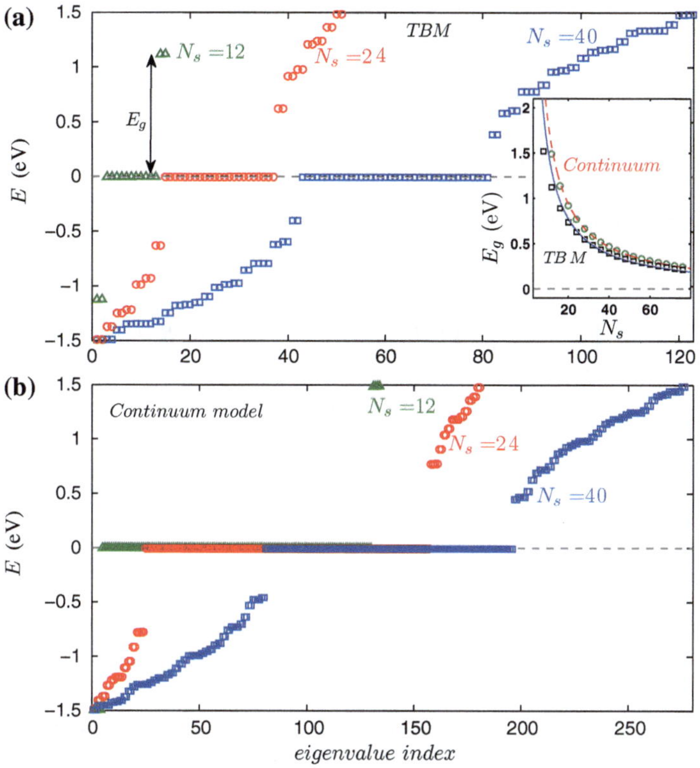

Fig. 4.9 Energy levels of a zigzag triangular graphene dot as a function of the eigenvalue index obtained by **a** the TBM and **b** the continuum model for three different sizes of the dot with $N_s = N_{edge} = 12, 24, 40$ having respectively surface area $S = 4.42, 16.37, 44.03\,\text{nm}^2$. The inset in panel (**a**) shows the energy gap obtained from both TBM (*black squares*) and continuum model (*green circles*). Reprinted from [13]

where $l_B = \sqrt{\hbar c/eB}$ is the magnetic length and c is the speed of light. Using the ansatz $\chi_A(r) = r^m \exp(-r^2/4l_B^2)\xi(r^2)$ one gets

$$\left[\tilde{r}\partial_{\tilde{r}}^2 + \left(m + 1 - \frac{\tilde{r}}{2l_B^2}\right)\partial_{\tilde{r}} + \frac{k^2 l_B^2 - 2(m+1)}{4l_B^2}\right]\xi_A(\tilde{r}) = 0, \qquad (4.13)$$

with $\tilde{r} = r^2$ and $\xi_A(\tilde{r}) = aL(k^2l_B^2/2 - (m+1), m, \tilde{r}/2l_B^2)$, where $L(k^2l_B^2/2 - (m+1), m, \tilde{r}/2l_B^2)$ is the generalized Laguerre polynomial and a is a normalization constant. (4.13) is a differential equation with a solution given by the confluent hypergeometric function of the first and second kind, with only the first kind having a physical meaning. The solution of (4.12) can be written

$$\chi_A(r) = ae^{im\phi}r^m e^{-r^2/4l_B^2} L(k^2l_B^2/2 - (m+1), m, r^2/2l_B^2), \qquad (4.14)$$

and for the second component

$$\chi_B(r) = aie^{i(m+1)\phi} r^m e^{-r^2/4l_B^2} \frac{r}{kl_B^2} L(k^2 l_B^2/2 - (m+2), m+1, r^2/2l_B^2) \\ + L(k^2 l_B^2/2 - (m+1), m, r^2/2l_B^2). \tag{4.15}$$

The infinite-mass boundary condition given by (4.6) gives an equation for allowed values of k

$$\left(1 - \tau \frac{k}{Rl_B^2}\right) L(k^2 l_B^2/2 - (m+1), m, R^2/2l_B^2) \\ + L(k^2 l_B^2/2 - (m+2), m+1, R^2/2l_B^2) = 0. \tag{4.16}$$

Taking the limit $R/l_B \to 0$ and solving for $k_m^2 l_B$ satisfying (4.16), one can retrieve the dependence of the energy on the magnetic field of a relativistic massless particle:

$$E_m = \hbar v_F k_m = \pm\sqrt{2e\hbar B(m+1)}. \tag{4.17}$$

In Fig. 4.10, we show the energy spectrum as a function of the magnetic field B of a quantum dot with $R = 70$ nm. In contrast with semiconductor quantum dots, the energy levels are not equidistant. The formation of Landau levels according to (4.17) for higher magnetic fields is visible. The zero Landau level is formed by states with quantum number $\tau = -1$ and $E > 0$, and those with $\tau = +1$ and $E < 0$. We note that the results obtained from (4.17), in particular the decrease of energy with the increasing magnetic field, were compared with the results of the Coulomb blockade transport spectroscopy showing qualitative agreement between theory and experiment [8, 26].

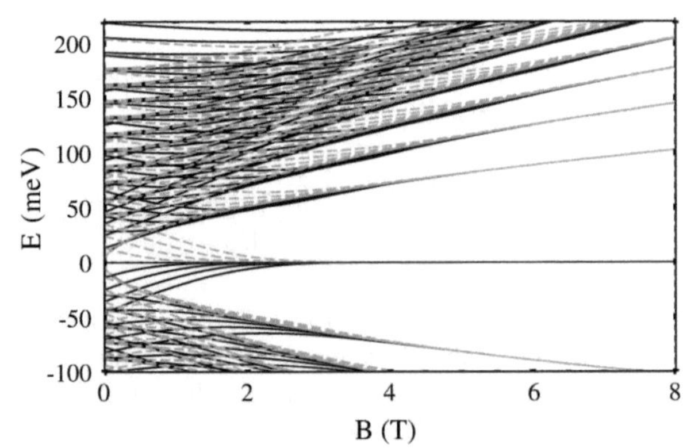

Fig. 4.10 Energy spectrum of a quantum dot with $R = 70$ nm. The formation of the lowest Landau levels can be seen as predicted by (4.17). Energy states for $\tau = +1$ are drawn with *solid lines*, those for $\tau = -1$ with *dashed lines*. Reprinted from [8]

In quantum dots described by the TB model, the magnetic field is incorporated using the Peierls substitution. The TB Hamiltonian can be written as

$$H_{TMF} = t \sum_{\langle i,l \rangle,\sigma} e^{i\phi_{ij}} c^{\dagger}_{i\sigma} c_{l\sigma}, \tag{4.18}$$

where for the symmetric gauge $\mathbf{A} = (B_z/2)(-y, x, 0)$, $\phi_{ij} = 2\pi e/h \int_{r_i}^{r_j} \mathbf{A}d\mathbf{l} = (B_z/2)(x_i y_j - y_i x_j)$. The magnetic field perpendicular to the dot plane can be expressed by the magnetic flux threading the area of a single benzene ring, $\phi = BS_0$, with $S_0 = 0.0524\,\text{nm}^2$. The flux can be measured in the units of the flux quantum $\phi_0 = \hbar c/e$. For $\phi/\phi_0 = 1$, there is exactly one magnetic flux quantum threading each benzene ring of the graphene quantum dot.

Zhang et al. studied the magnetic field dependence of the energy spectrum of hexagonal dots with zigzag and armchair edges [19]. The energy spectra corresponding to a hexagonal dot with $N = 864$ atoms with zigzag edges and $N_{edge} = 12$ atoms on each edge, are shown in Fig. 4.11. At high magnetic fields Fig. 4.11a shows the formation of the Hofstadter butterfly, a fractal energy spectrum in magnetic fields [27, 28]. For smaller magnetic fields, Landau levels form according to formula given by (4.17) which is observed in Fig. 4.11b. Energy levels approach the zeroth Landau level at zero energy in pairs, one from the valence and one from the conduction band, as seen in Fig. 4.11c. For $\phi/\phi_0 = 1/2N_{edge}$ the degeneracy of the zeroth Landau level is maximal, equal to $2N_{edge}$. For larger magnetic fields, the zeroth Landau level splits. In Fig. 4.11d the DOS at the Dirac point, $E = 0$, is shown. The number of energy levels at $E = 0$ decreases approximately inversely with the magnetic flux ϕ/ϕ_0. The results for hexagonal dot with armchair edges in an external magnetic field were similar to that with zigzag edges. The only difference is observed for small magnetic fields and is related to the distinct behavior of edge states present in GQDs with zigzag edges.

Graphene quantum dots and rings with different shapes were also studied [13, 15, 20, 23, 24]. Independently of the shape, all energy spectra converge to the Landau levels of graphene as the magnetic field increases. This agrees with our intuition, for large magnetic fields the confinement is dominated by the magnetic field and the influence of the shape and edge is suppressed. On the other hand, the energy gap in an external magnetic field behaves differently for triangular and hexagonal dots. In the case of the hexagonal dot the energy gap closes quickly, while for triangles there is an almost linear dependence with the magnetic flux [13].

4.2 Spin-Orbit Coupling in Graphene Quantum Dots

Until now, we described the properties of graphene quantum dots using only p_z orbitals. As discussed in the introduction, electronic configuration of carbon atom is $1s^2 2s^2 2p^2$, where the s, p_x and p_y orbitals hybridize to form sp^2 bonds (sigma bonds), responsible for the honeycomb lattice structure and mechanical properties

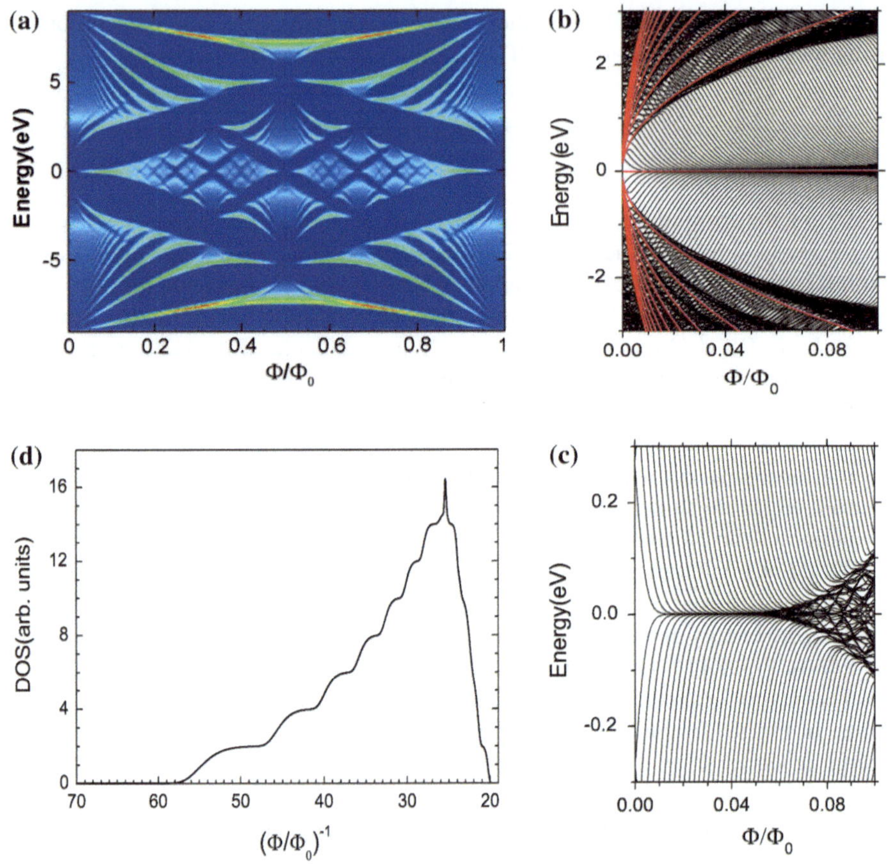

Fig. 4.11 **a** The DOS and energy spectrum of the $N_{edge} = 12$ ZGQD in a magnetic field. A Gauss function with a broadening factor of 0.1 eV to smoothen the discontinuous energy spectra was used. **b** and **c** The magnetic energy level fan near the Dirac point, i.e., the zero-energy point. The *red lines* in (**b**) correspond to the Landau level of two-dimensional graphene given by (4.17). **d** the DOS at the Dirac point, as a function of the inverse flux ϕ/ϕ_0, where a Gauss function with a small broadening factor of 0.01 meV was used. Reprinted from [19]

of graphene. Since p_z orbitals are perpendicular to all neighboring sigma orbitals, as a first approximation they were considered decoupled. However, as discussed by Dresselhaus [29], inclusion of the spin-orbit coupling requires coupling of p_z orbitals with sigma orbitals and d-orbitals [30]. We describe below the spin-orbit coupling within the four-band tight-binding model taking into account the mixing between s, p_x, p_y, and p_z orbitals. We compare the four-band tight-binding model with an effective spin-dependent p_z orbital Kane-Mele model [31] used to study the quantum spin Hall effect in graphene nanoribbons.

4.2.1 Four-Band Tight-Binding Model

The four-band extension of the single-band TB Hamiltonian given in (4.1) can be written as:

$$H_{TB} = \sum_{\langle i,l \rangle} \sum_{\mu_i,\mu_l,\sigma} t_{\mu_i,\mu_l} c^{\dagger}_{i\mu_i\sigma} c_{l\mu_l\sigma}, \tag{4.19}$$

where i and l are site indices, and μ labels one of the four orbitals, s, p_x, p_y, or p_z. The tight-binding parameters for neighboring atoms can be conveniently expanded as a function of four nonzero and linearly independent hopping parameters, t^{σ}_{ss}, t^{π}_{pp}, t^{σ}_{sp}, t^{σ}_{pp}, illustrated in Fig. 4.12. Examples of this expansion are given in Fig. 4.13, where the tight-binding parameters t_{μ_1,μ_2} are given as a function of a unit vector n starting from one site and ending on the second one. We see immediately that $t_{sp_z} = t_{p_x p_z} = t_{p_y p_z} = 0$ as a result of symmetry, i.e., p_z orbitals do not couple to sigma bonds, unless the spin-orbit coupling is included.

In the four-orbital tight-binding model we must specify passivation of the edges of graphene nanostructures. The edge atoms have only two carbon neighbors which leaves one of the three sp^2 bonds as a dangling bond. The dangling bonds can be passivated by hydrogen atoms which do not contribute significantly to the p_z electrons of graphene. Thus, unlike in the single-orbital tight-binding model, a proper four-orbital treatment of graphene nanostructures must include hydrogen atoms attached to each edge atom, keeping the structure of the sp^2 bonds intact at the edges. An example of a hydrogen passivated graphene quantum dot is shown in Fig. 4.14, where the structure has $N = 97$ carbon atoms (dark color) and 27 hydrogen atoms (light color). Electronic properties of this structure will be studied in detail in the following section. All the tight-binding parameters including the Carbon–Carbon (C–C) and Carbon–Hydrogen (C–H) hopping matrix elements, overlap matrix elements and on-site energies of carbon and hydrogen atoms, are given in Table 4.1.

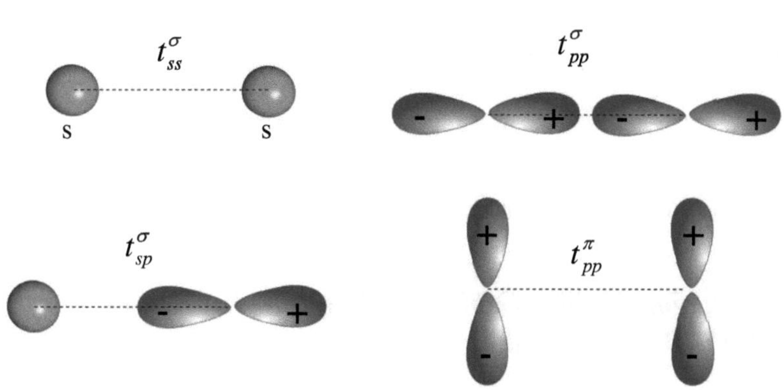

Fig. 4.12 Schematic illustration of s and p_x, p_y, p_z orbitals and their contribution to tunneling matrix elements

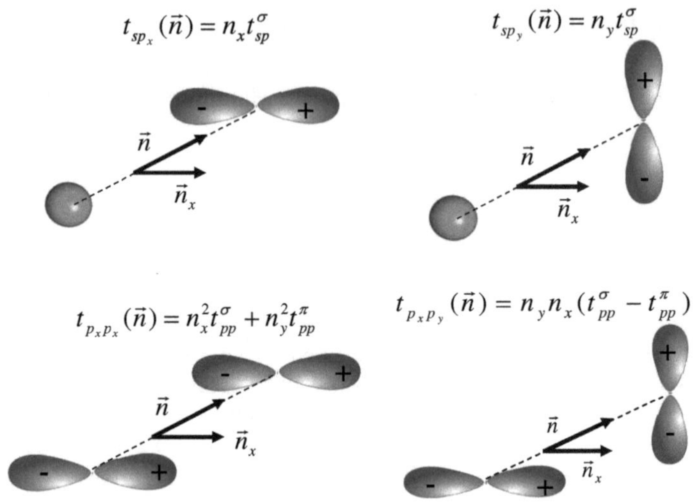

Fig. 4.13 Decomposition of tunneling matrix elements into contributions from different atomic orbitals

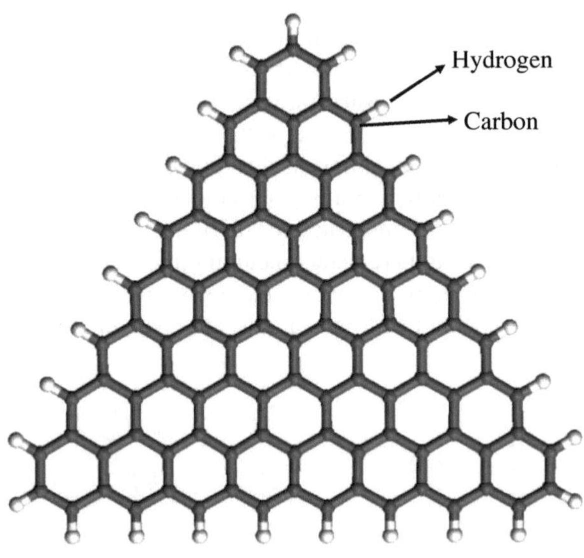

Fig. 4.14 Triangular graphene quantum dot with hydrogen passivated edges

4.2.2 Inclusion of Spin-Orbit Coupling into Four-Band Tight-Binding Model

As an electron moves in the electrostatic field of the nucleus, according to special relativity, a magnetic field appears in the reference frame of the electron. The resulting

Table 4.1 Tight-binding parameters including the Carbon–Carbon (C–C) and Carbon–Hydrogen (C–H) hopping matrix elements, overlap matrix elements and on-site energies of Carbon and Hydrogen atoms

Hopping matrix elements				
	t_{ss}^{σ}	t_{sp}^{σ}	t_{pp}^{σ}	t_{pp}^{π}
C–C	−6.769	5.58	5.037	−2.8
C–H	−5.4	5.8	–	–
Overlap matrix elements				
	$S_{ss}^{\sigma} = 0.212$	$S_{sp}^{\sigma} = -0.102$	$S_{pp}^{\sigma} = -0.146$	$S_{pp}^{\pi} = 0.129$
On-site energies				
	$\varepsilon_s^C = -8.868$	$\varepsilon_p^C = 0.0$	$\varepsilon_s^H = -1.2$	

Parameters taken from [32, 33]

magnetic field interacts with the electron spin, giving rise to the spin-orbit interaction. The spin-orbit coupling scales with the atomic number of the atom, thus its effect is weaker for the carbon atom compared to heavier atoms such as gallium or arsenic. Nevertheless, the spin-orbit coupling in a graphene nanostructure may lead to the spin Hall effect and control the conversion of the photon angular momentum to the electron spin as in semiconductor quantum dots.

In the tight-binding model the spin-orbit Hamiltonian is usually parameterized by the $\mathbf{L} \cdot \mathbf{S}$ coupling on each atom, with the *SO* Hamiltonian written as [29, 34, 35]:

$$H_{SO} = \sum_{i,l} \lambda_l \mathbf{L_i} \cdot \mathbf{S_i}, \tag{4.20}$$

where i is the site index and λ_l is the angular-momentum-resolved atomic spin-orbit coupling strength with $l = \{s, p, d, \ldots\}$. In our four-orbital tight-binding model, the s orbitals do not contribute to the spin-orbit coupling since their angular momentum is zero, the summation over l is restricted to p orbitals only. In order to calculate the spin-orbit matrix elements, we can rewrite (4.20) in terms of angular momentum ladder operators:

$$H_{SO} = \sum_i \lambda \left(\frac{L_+^i S_-^i + L_-^i S_+^i}{2} + L_z^i S_z^i \right). \tag{4.21}$$

Next, we express orbitals p_x, p_y, p_z in terms of spherical harmonics $|l, m_l\rangle$ which are eigenstates of the L^2 and L_z operators:

$$\begin{aligned} |p_x\rangle &= \frac{1}{\sqrt{2}} (|1, 1\rangle + |1, -1\rangle) \\ |p_y\rangle &= -\frac{i}{\sqrt{2}} (|1, 1\rangle - |1, -1\rangle) \\ |p_z\rangle &= |1, 0\rangle. \end{aligned} \tag{4.22}$$

For a given site i, the Hamiltonian matrix elements can then be calculated using (4.21) and (4.22):

$$
\begin{array}{c|cccccc}
 & |p_x,\uparrow\rangle & |p_y,\uparrow\rangle & |p_z,\uparrow\rangle & |p_x,\downarrow\rangle & |p_y,\downarrow\rangle & |p_z,\downarrow\rangle \\
\langle p_x,\uparrow| & 0 & \frac{-\lambda i}{2} & 0 & 0 & 0 & \frac{\lambda}{2} \\
\langle p_y,\uparrow| & \frac{\lambda i}{2} & 0 & 0 & 0 & 0 & \frac{-\lambda i}{2} \\
\langle p_z,\uparrow| & 0 & 0 & 0 & \frac{\lambda}{2} & \frac{-\lambda i}{2} & 0 \\
\langle p_x,\downarrow| & 0 & 0 & \frac{\lambda}{2} & 0 & \frac{\lambda i}{2} & 0 \\
\langle p_y,\downarrow| & 0 & 0 & \frac{\lambda i}{2} & \frac{-\lambda i}{2} & 0 & 0 \\
\langle p_z,\downarrow| & \frac{\lambda}{2} & \frac{\lambda i}{2} & 0 & 0 & 0 & 0
\end{array}
\quad . \tag{4.23}
$$

Note that with the spin-orbit interaction, the p_z orbital now couples to p_x and p_y orbitals. Using (4.23) together with (4.19) we can construct the full tight-binding Hamiltonian matrix for a graphene nanostructure of arbitrary shape and study the effect of the spin-orbit coupling on its electronic properties. In Sect. 4.3 we will apply the four-band Hamiltonian to understand the energy spectrum and orbitals of a triangular quantum dot with zigzag edges.

4.2.3 Kane-Mele Hamiltonian and Quantum Spin Hall Effect in Nanoribbons

We have seen in the previous section that in order to study spin-orbit interactions, one needs to expand the tight-binding basis set to include all four valence orbitals. Furthermore, recent results [30] show that the contribution of d orbitals can also be important, which increases the number of orbitals and the size of the TB Hamiltonian matrix. In 2005, Kane and Mele [31] proposed an effective spin-orbit tight-binding Hamiltonian which involves only p_z orbitals, and is given by:

$$H_{SO} = it_{SO} \sum_{\langle\langle i,l\rangle\rangle\sigma} \nu_{il}\sigma c^{\dagger}_{i\sigma} c_{l\sigma}, \tag{4.24}$$

where t_{SO} is an effective second neighbor spin-orbit hopping parameter. The second neighbors are connected by a spin-dependent amplitude and through $\nu_{il} = -\nu_{li} = \pm 1$. The value of ν_{il} depends on the orientation of the two neighbors: $\nu_{il} = +1(-1)$ if going from site i to reach site l the electron makes a left (right) turn. This procedure mimics the microscopic spin-orbit interaction where an electron in p_z orbital can first hop into a p_x (p_y) orbital through spin-orbit coupling, then hop onto the s orbital of a neighboring atom, which is followed by another first neighbor hopping into the p_y (p_x), finally ending up in the p_z orbital [30]. The result is an orientation- and spin-dependent effective second neighbor hopping between p_z orbitals. Note that, although (4.24) does not conserve the total spin (as expected from a $\mathbf{L} \cdot \mathbf{S}$ coupling),

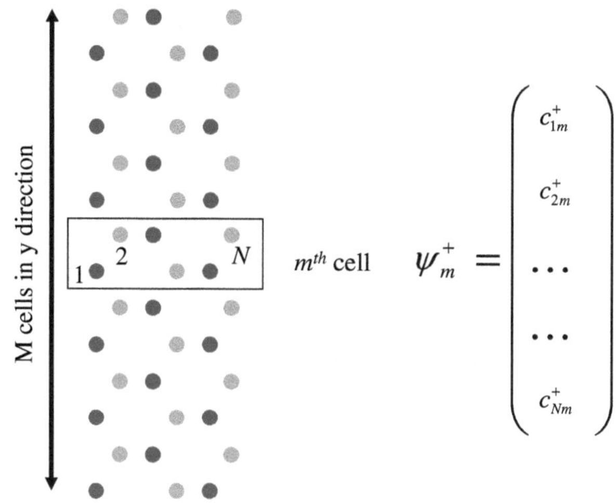

Fig. 4.15 A schematic representation of a graphene nanoribbon. On the *right*, a creation operator on cell m containing N sites

it does conserve the projection of total spin, S_z. For spin-up electrons we have:

$$H_{SO}^{\uparrow} = it_{SO} \sum_{\langle\langle i,l \rangle\rangle} v_{il} c_{i\uparrow}^{\dagger} c_{l\uparrow}, \tag{4.25}$$

which is the complex conjugate of $H_{SO}^{\downarrow}$. Both Hamiltonians share the same eigenvalues and their eigenfunctions are also complex conjugates of each other, i.e. $\psi_{\uparrow} = \psi_{\downarrow}^*$. This is a special case of the Kramers degeneracy.

A remarkable consequence of the spin-orbit coupling, the quantum spin Hall effect, arises in graphene nanoribbon structures as demonstrated by Kane and Mele [31]. Let us consider the graphene ribbon shown in Fig. 4.15, periodic in the y-direction with M cells denoted with index m. Each cell has N sites, denoted by index n. N determines the width of the ribbon. The single-orbital tight-binding Hamiltonian can be written as:

$$H = \sum_{n_1 m_1 n_2 m_2 > \sigma} \tau_{n_1 m_1 n_2 m_2 \sigma} c_{n_1 m_1 \sigma}^{\dagger} c_{n_2 m_2 \sigma}, \tag{4.26}$$

where τ includes nearest-neighbor hoppings and spin-orbit second-nearest-neighbor hoppings. We can simplify the problem by defining a "cell creation" operator ψ_m, shown in Fig. 4.15 [36]. The tight-binding Hamiltonian then can be rewritten in terms of operators ψ_m as

$$H = \sum_m \psi_m^{\dagger} U \psi_m + \left(\psi_m^{\dagger} T \psi_{m+1} + \text{h.c.} \right), \tag{4.27}$$

where the matrix U describes the hopping terms between sites within the cell m, whereas the matrix T describes the hopping terms between cells m and $m+1$. The U and T matrices can be generated by inspection. In the subspace of spin-up electrons they are given by:

$$U_{\uparrow} = \begin{pmatrix} 0 & t & -it_{SO} & 0 & 0 & 0 & \dots \\ t & 0 & t & -it_{SO} & 0 & 0 & \\ it_{SO} & t & 0 & t & it_{SO} & 0 & \\ 0 & it_{SO} & t & 0 & t & it_{SO} & \\ 0 & 0 & -it_{SO} & t & 0 & t & \\ 0 & 0 & 0 & -it_{SO} & t & 0 & \\ \dots & & & & & & \dots \end{pmatrix}, \tag{4.28}$$

and

$$T_{\uparrow} = \begin{pmatrix} it_{SO} & 0 & 0 & 0 & 0 & 0 & \dots \\ t & -it_{SO} & 0 & it_{SO} & 0 & 0 & \\ -it_{SO} & 0 & it_{SO} & t & -it_{SO} & 0 & \\ 0 & 0 & 0 & -it_{SO} & 0 & 0 & \\ 0 & 0 & 0 & 0 & it_{SO} & 0 & \\ 0 & 0 & 0 & it_{SO} & t & -it_{SO} & \\ \dots & & & & & & \dots \end{pmatrix}. \tag{4.29}$$

The periodicity of the ribbon in y-direction allows us to write

$$\psi_m = \psi_{m+M} = \frac{1}{\sqrt{M}} \sum_{k=0}^{M-1} \psi_k e^{-i2\pi km/M}, \tag{4.30}$$

where ψ_k is the Fourier transform of ψ_m. Finally, the Hamiltonian becomes

$$H = \sum_k \psi_k^{\dagger} \left(U + \left(T e^{i2\pi k/M} + \text{h.c.} \right) \right) \psi_k, \tag{4.31}$$

which is diagonal in k. The band structure of graphene ribbon for a given N, M, and k, can now be easily calculated by diagonalizing $U_{\uparrow} + \left(T_{\uparrow} e^{i2\pi k/M} + \text{h.c.}\right)$.

Figure 4.16 shows the band structure of a graphene ribbon with $M = 1{,}000$ and $N = 56$ as a function of $k_y = 2\pi k/L = 2\pi k/Ma\sqrt{3}$, with spin-orbit coupling parameter set to zero, $t_{SO} = 0$. L is the length of the ribbon. Due to the presence of zigzag edges, a doubly degenerate zero-energy band appears at the Fermi level [37]. The states corresponding to the zero-energy band are localized at both edges of the ribbon, as shown in the inset of Fig. 4.16 for a particular k_y indicated on the zero-energy band. When spin-orbit coupling is present (here we take $t_{SO} = 0.03t$), the doubly degenerate zero-energy band splits into two bands, each localized on one edge (see Fig. 4.17, left-hand side). As the slopes of the two bands are opposite to each other, they correspond to waves travelling in opposite directions. More strikingly,

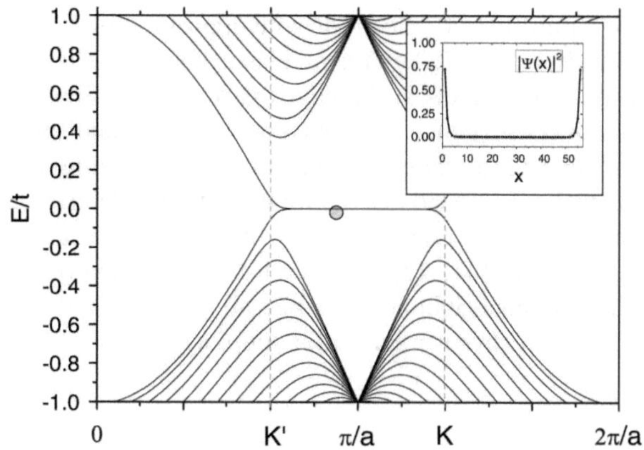

Fig. 4.16 Energy bands $E(i, k_y)$ of graphene ribbon with $M = 1{,}000$ and $N = 56$ and no SO coupling

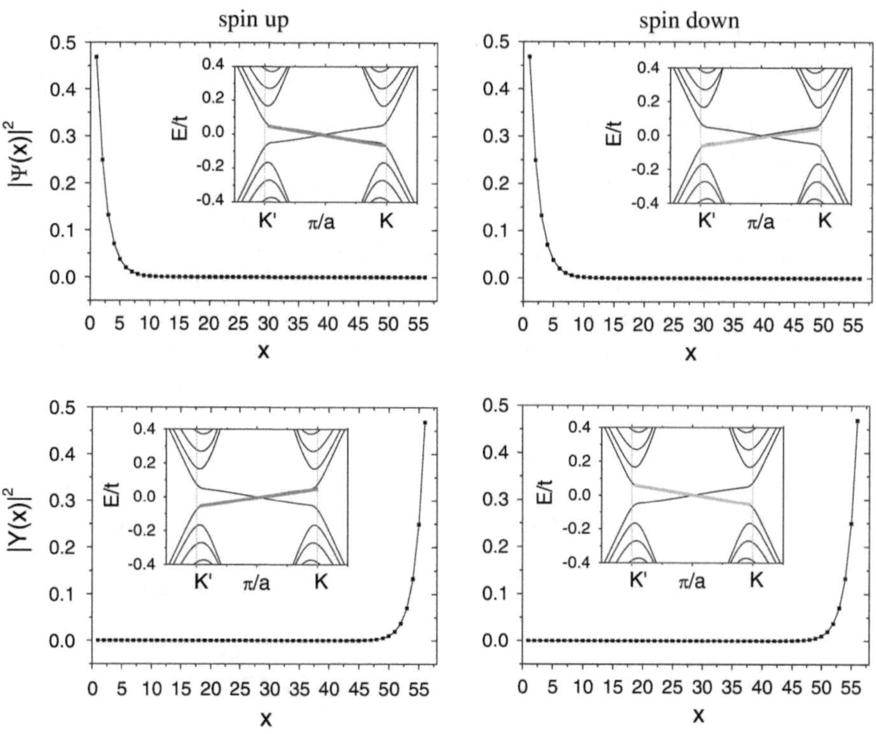

Fig. 4.17 Energy bands $E(i, k_y)$ of graphene ribbon with $M = 1{,}000$ and $N = 56$ with SO coupling

the edge states are "spin-filtered": electrons with opposite spin propagate in opposite directions, as shown in Fig. 4.17. As a result, elastic backscattering by random (non-magnetic) impurities and defects at the edges is forbidden. The spin filtered edge states have important consequences for transport of charge and spin [38]. Although the ideal bulk graphene has a relatively small spin-orbit splitting, the effective spin-orbit coupling strength can depend heavily on the external electric field [39, 40], the curvature present in graphene due to strain or impurities, and electron occupation in graphene nanostructures [41, 42]. Moreover, the prediction of the quantum spin Hall effect in graphene nanoribbons described above generated interest in finding alternative materials with stronger spin-orbit coupling [38].

4.3 Triangular Graphene Quantum Dots with Zigzag Edges

4.3.1 Energy Spectrum

In Sect. 4.1.1, we have shown numerical evidence that the energy spectrum of triangular graphene quantum dots (TGQD) with zigzag edges is characterized by the existence of a degenerate energy shell at the Fermi level. TGQDs are an example of graphene nanostructures with broken sublattice symmetry. Here, we carry out a detailed analysis of single-particle properties of TGQD.

Each TGQD can be characterized by the number of atoms on one edge, N_{edge}, and the total number of atoms $N = N_{edge}^2 + 4N_{edge} + 1$ is expressed by the number of atoms at the edge. There are N_A and N_B atoms corresponding to sublattice A and B. The difference between the number of atoms of types A and B is proportional to the number of atoms at one edge, $|N_A - N_B| = N_{edge} - 1$. This feature is critical for the explanation of the existence of the degenerate energy shell in TB energy spectra in Sect. 2.2.2.

We now relate the number of states in the degenerate shell to the number of atoms at the edge. In Fig. 4.18 we show the TB energy spectra of two TGQDs

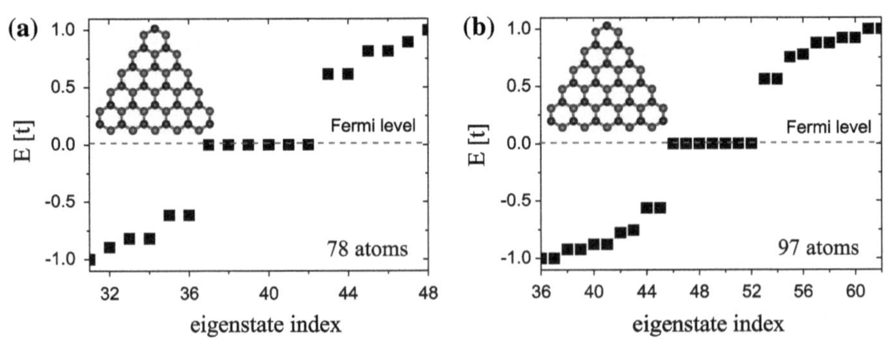

Fig. 4.18 TB energy spectra of TGQDs consisting of (**a**) $N = 78$ atoms ($N_{edge} = 7$) and $N = 97$ atoms ($N_{edge} = 8$). There are **a** $N_{deg} = 6$ and **b** $N_{deg} = 7$ degenerate states at the Fermi level

with different sizes. Figure 4.18a corresponds to the structure consisting of $N = 78$ atoms, or $N_{edge} = 7$, and Fig. 4.18b to the structure consisting of $N = 97$ atoms or $N_{edge} = 8$. From numerical diagonalization of the TB Hamiltonian we find $N_{deg} = 6$ and $N_{deg} = 7$ degenerate states at the Fermi level, respectively. The number of degenerate states N_{deg} in these TGQDs is related to the number of edge atoms as $N_{deg} = N_{edge} - 1 = N_A - N_B$. In the next subsection we prove that this is a general rule for all TGQDs, by increasing the size of triangles the degeneracy of the zero-energy shell increases and can be made macroscopic.

We now examine the electronic probability densities of the degenerate zero-energy levels. We focus on the structure with $N = 97$ atoms and the energy spectrum shown in Fig. 4.18b. There are $N_{deg} = 7$ degenerate energy levels. Due to a perfect degeneracy of these states, arbitrary linear combinations of seven eigenfunctions can be constructed. Thus, in order to preserve the triangular symmetry of eigenstates, the degeneracy is slightly removed by applying a very small random energy shift on each atomic site. The seven-fold degenerate shell is split into two doubly degenerate and three non-degenerate states, with electronic probability densities shown in Fig. 4.19(a–e). The radius of circles is proportional to the electronic probability density on an atomic site. In the case of the doubly degenerate state, the sum of electronic densities corresponding to these two states is plotted. For single non-degenerate states, the probability density is multiplied by a factor of two compared with doubly degenerate states. Five of these states (Fig. 4.19a–c), are strongly localized at the edges. Last two states, shown in Fig. 4.19d, e, fill the center of the triangle and the center of edges avoiding corners. While these two states contribute to the electronic probability density in the center of the triangle, when all densities are added up it is a small contribution in comparison to the electronic probability density localized on edges. This is shown in Fig. 4.19f, where the total charge density of the zero-energy shell is plotted. Proportions between Figs. 4.19a–e and 4.19f are not maintained. We note that all states are localized only on the sublattice A, as indicated by red color.

4.3.2 Analytical Solution for Zero-Energy States

Following [43] we find zero-energy solutions of the TB Hamiltonian given by (4.1) and show that the degeneracy of the zero-energy shell is proportional to the number of atoms at the edge of the TGQD. The zero-energy shell corresponds to solutions of a singular eigenvalue problem written as

$$H_{TB}\Psi = 0. \tag{4.32}$$

In order to clearly explain our methodology, we first write (4.32) for an arbitrary B-type site surrounded by three A-type sites, shown in Fig. 4.20a, as it takes place

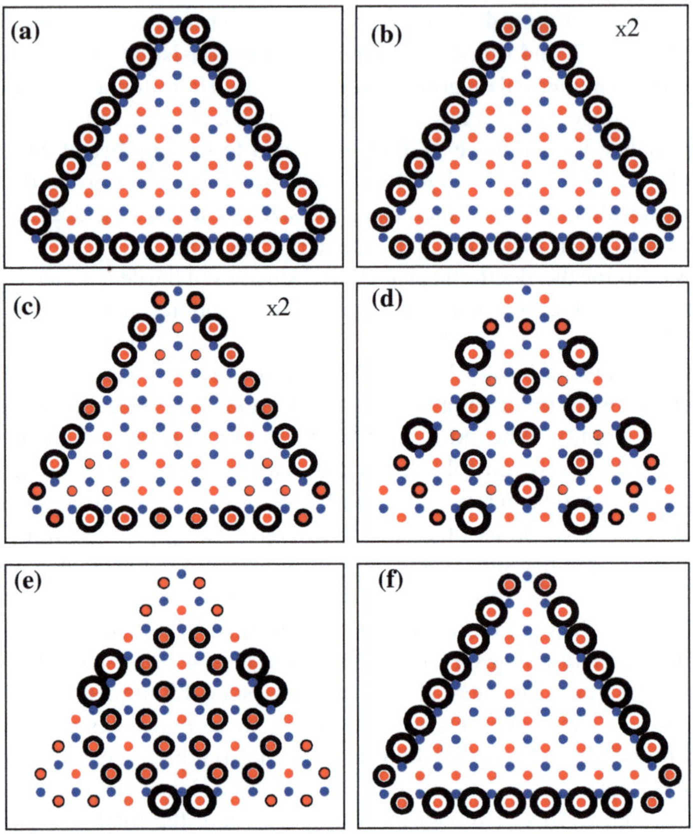

Fig. 4.19 **a–e** Electronic densities of $N_{deg} = 7$ degenerate energy levels with $E = 0$ for structure consisting of $N = 97$ atoms. **a–c** Five states strongly localized on edges. **d, e** Two states localized in the *center* of the *triangle*. **f** The total charge density of the zero-energy shell. All states are localized only on A sublattice, indicated by *red* color. A radius of *black circles* is proportional to the electronic density on an atomic site

for every bulk site in a honeycomb lattice; edge sites have only two neighbors. In this case, the wavefunction written in a basis of p_z orbitals ϕ_z has the following form

$$\Psi = b_i \phi_z^i + b_j \phi_z^j + b_k \phi_z^k + b_l \phi_z^l, \tag{4.33}$$

where b_i, b_j, b_k, b_l are expansion coefficients. Using (4.32) and projecting onto ϕ_z^i we get

$$\langle \phi_z^i | H_{TB} | \Psi \rangle = 0 \cdot b_i + t \cdot b_j + t \cdot b_k + t \cdot b_l = 0, \tag{4.34}$$

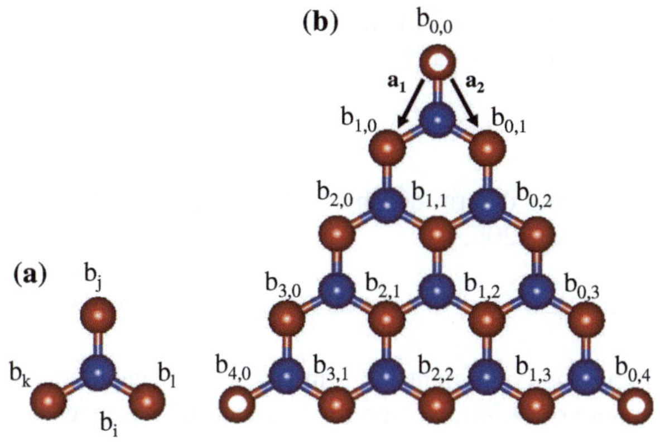

Fig. 4.20 **a** An arbitrary ith B-type site (*blue circle*) surrounded by three A-type sites, jth, kth, and lth, (*red circles*). **b** TGQD with $N_{edge} = 3$ atoms on one edge. Above each A-type atom are corresponding coefficients. *Open circles* indicate auxiliary A-type atoms in the three corners, which will help to introduce three boundary conditions. For zero-energy states all coefficients can be expressed as superpositions of coefficients from the one edge, the *left* edge of atoms in our case

where we use expressions $\langle\phi_z^i|H_{TB}|\phi_z^j\rangle = \langle\phi_z^i|H_{TB}|\phi_z^k\rangle = \langle\phi_z^i|H_{TB}|\phi_z^l\rangle = t$ and $\langle\phi_z^i|H_{TB}|\phi_z^i\rangle = 0$. Finally, (4.34) is written as

$$b_j + b_k + b_l = 0. \tag{4.35}$$

This brings us to important conclusions regarding eigenfunctions corresponding to zero-energy states: (i) there is no coupling between the two sublattices—coefficients b_j, b_k, b_l belong to one sublattice (note that this can be seen also from (2.16) for infinite graphene), (ii) the sum of coefficients around each site must vanish. These facts are valid for both sublattices. We now apply this knowledge to TGQDs.

We consider TGQD with $N = 22$ atoms and $N_{edge} = 3$ plotted in Fig. 4.20b. Let us first focus on the sublattice labeled by A, represented by red circles. The position of each atom is defined by a vector $\mathbf{R} = n\mathbf{a}_1 + m\mathbf{a}_2$, where n, m are two integer numbers, with $0 \leq n, m \leq N_{edge} + 1$, and $b_{n,m}$ are corresponding expansion coefficients in the basis of p_z orbitals. The structure has three auxiliary atoms attached with the coefficients $b_{0,0}, b_{0,4}, b_{4,0}$, which will later define appropriate boundary conditions. The auxiliary atoms are indicated by open circles in Fig. 4.20b. We start from the top of TGQD. For the first three coefficients $b_{0,0}, b_{1,0}, b_{0,1}$, (4.35) gives

$$b_{0,1} = -(b_{0,0} + b_{1,0}). \tag{4.36}$$

Thus, the coefficient $b_{0,1}$ is expressed using two coefficients from the left edge. We can take the next two coefficients from the left edge, $b_{1,0}$ and $b_{2,0}$, and obtain the coefficient $b_{1,1}$, which is written as

$$b_{1,1} = -(b_{1,0} + b_{2,0}). \tag{4.37}$$

The similarity of (4.36) and (4.37) leads us to a general expression for coefficients $b_{n,1}$, with $n = 1, 2, 3, 4$ which can be written as

$$b_{n,1} = -(b_{n,0} + b_{n+1,0}). \tag{4.38}$$

Additionally, coefficients determined by (4.36) and (4.37) allow to determine the coefficient $b_{0,2}$, see Fig. 4.20b,

$$b_{0,2} = -(b_{0,1} + b_{1,1}) = (b_{0,0} + 2b_{1,0} + b_{2,0}). \tag{4.39}$$

Thus, in general, having coefficients $b_{n,1}$ and $b_{n+1,1}$ one can determine coefficients $b_{n,2}$, which with help of (4.38), written for $b_{n,1}$ and $b_{n+1,1}$, gives

$$b_{n,2} = -(b_{n,1} + b_{n+1,1}) = (b_{n,0} + 2b_{n+1,0} + b_{n+2,0}). \tag{4.40}$$

But coefficients $b_{n,2}$ and $b_{n+1,2}$ determine the coefficient $b_{n,3}$, see for example the coefficient $b_{0,3}$ in Fig. 4.20b. In order to write an expression for coefficients $b_{n,3}$, (4.40) for $b_{n,2}$ and $b_{n+1,2}$ has to be combined to give

$$b_{n,3} = (b_{n,0} + 3b_{n+1,0} + 3b_{n+2,0} + b_{n+3,0}). \tag{4.41}$$

Going farther in this way, coefficients $b_{n,4}$ can be obtained from $b_{n,3}$ and $b_{n+1,3}$. For example the coefficient $b_{0,4}$ from coefficients $b_{0,3}$ and $b_{1,3}$, see Fig. 4.20b, which can be written as

$$b_{0,4} = b_{0,0} + 4b_{1,0} + 6b_{2,0} + 4b_{3,0} + b_{4,0}. \tag{4.42}$$

Looking at (4.38), (4.40), and (4.41) one can see that all coefficients for the A-type atoms in TGQD from Fig. 4.20b are determined by coefficients from the left edge, $b_{n,0}$. One can also see that numbers standing next to coefficients are identical to those in the Pascal triangle [44]. Thus, these coefficients can be written in a compact form using binomial coefficients

$$b_{n,m} = (-1)^m \sum_{k=0}^{m} \binom{m}{k} b_{n+k,0}. \tag{4.43}$$

Here, it is important to emphasize that the only unknowns are the $N_{edge} + 2$ coefficients, $b_{n,0}$'s, from the left edge; the rest are expressed as their superpositions, as seen from (4.43). In addition, we must use the boundary conditions: the wave function has to vanish on three auxiliary atoms in each corner, see Fig. 4.20b. This gives three boundary conditions, for TGQD from Fig. 4.20b ($b_{0,0} = b_{4,0} = b_{0,4} = 0$), or for arbitrary-size triangle ($b_{0,0} = b_{N_{edge}+1,0} = b_{0,N_{edge}+1} = 0$), reducing the number of independent coefficients to $N_{edge} - 1$. The number of linearly independent coefficients corresponds to the maximum number of linearly independent vectors and

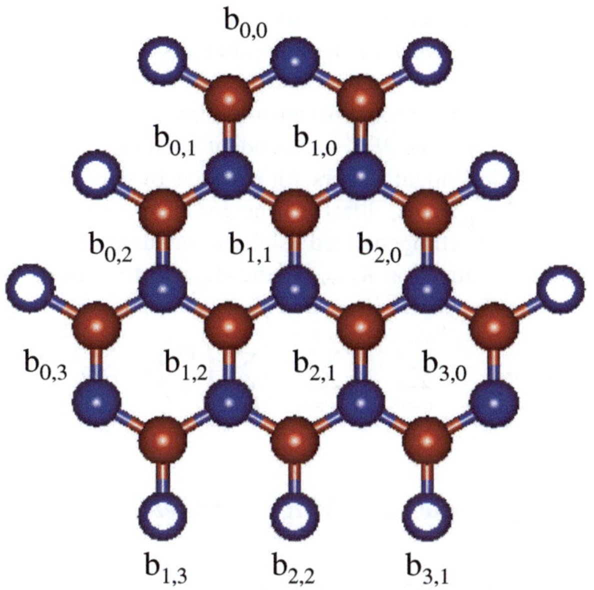

Fig. 4.21 TGQD from Fig. 4.20. Above each B-type atom (indicated by *blue circles*) are corresponding coefficients. We only left coefficients corresponding to auxiliary B-type atoms from the bottom. For zero-energy states, a coefficient from lower left corner ($b_{0,1}$) determines all other coefficients. Introducing four boundary conditions from auxiliary atoms, we obtain only trivial solution

determines the dimension of the degenerate zero-energy shell $N_{deg} = N_{edge} - 1$. This result confirms our previous numerical calculations, shown in Fig. 4.18.

The same analysis of zero-energy states can be done for B-type atoms indicated by blue circles. In this case it is convenient to include some of the boundary conditions at the beginning as shown in Fig. 4.21, where we only keep coefficients belonging to auxiliary atoms from the bottom edge. As a consequence, the coefficient $b_{0,0}$ determines all other coefficients in the triangle. Since there are three auxiliary atoms (equivalently three boundary conditions) but only one independent coefficient, we can not obtain any nontrivial solution. Hence, zero-energy states can only consist of coefficients of one type of atoms—these lying on the edges (A-type atoms). A general form for the eigenvectors for zero-energy states in the triangle can be written as

$$\Psi_{E=0} = \sum_{n=0}^{N_{edge}+1} \sum_{m=0}^{N_{edge}+1-n} \left[(-1)^m \sum_{k=0}^{m} \binom{m}{k} b_{n+k,0} \right] \phi^A_{n,m}, \tag{4.44}$$

where N_{edge} is the number of atoms on one edge and $\phi^A_{n,m}$ is the p_z orbital on the A-type site (n, m). In this expression only $N_{edge} - 1$ coefficients corresponding to atoms from the one edge are independent. We can construct $N_{edge} - 1$ linear independent eigenvectors, which span the subspace of zero-energy states. Thus, the number of zero-energy states in the triangle is $N_{deg} = N_{edge} - 1$. This can be also related to the imbalance between the number of atoms belonging to each sublattice, $N_{deg} = N_A - N_B$.

Using the (4.44) we can now construct an orthonormal basis for zero-energy states. First, we make a choice for the N_{deg} independent coefficients $b_{n,0}$, from which we obtain N_{deg} linearly independent vectors, for instance, by choosing only one nonzero coefficient for each N_{deg} states, different one for each eigenvector. The resulting eigenvectors can then be orthogonalized using the standard Gram-Schmidt process. The last step is the normalization K_{norm} of the eigenvectors, using the expression

$$K_{norm} = \sum_{n=0}^{N_{edge}+1} \sum_{m=0}^{N_{edge}+1-n} \left| \sum_{k=0}^{m} \binom{m}{k} b_{n+k,0} \right|^2 .$$

4.3.3 Zero-Energy States in a Magnetic Field

The analysis of the zero-energy states can also be generalized to non-zero external magnetic fields [25]. In this case, the wave function coefficients given in the bracket in (4.44) become

$$b_{n,m}(\phi) = (-1)^m \sum_{k=0}^{m} \frac{1 - e^{2\pi i \binom{m}{k} \frac{\phi}{\phi_0}}}{1 - e^{2\pi i \frac{\phi}{\phi_0}}} e^{-i\varphi_{n+k}} b_{n+k,0}, \tag{4.45}$$

where $\phi_0 = \frac{hc}{e}$ is the magnetic flux quantum, $\phi = B_z S_0$ is the magnetic flux threading one benzene ring, $S_0 = 3\sqrt{3}a_0^2/2$ is the benzene ring area with $a_0 = 1.42\,\text{Å}$, and φ_{n+k} represents the phase corresponding to the path on the right edge connecting sites $\{n+k, 0\}$ and $\{n, m\}$ [25]. Note that (4.45) reduces to (4.43) when $\phi = 0$. Interestingly, (4.45) shows that the zero energy states in triangular graphene quantum dots survive in external magnetic fields, the only effect is the Zeeman splitting. The effect is similar to the appearance of the $n = 0$ Landau level in bulk graphene. When the cyclotron energy becomes comparable to the energy gap, the zero-energy shell and electron and hole states evolving toward the $n = 0$ Landau level overlap energetically, opening the possibility of manipulating strongly correlated electronic systems of the degenerate zero-energy shell [25].

4.3.4 Classification of States with Respect to Irreducible Representations of C_{3v} Symmetry Group

Following [45] we will apply the group theory to classify electronic states of triangular graphene quantum dots according to irreducible representations of the C_{3v}

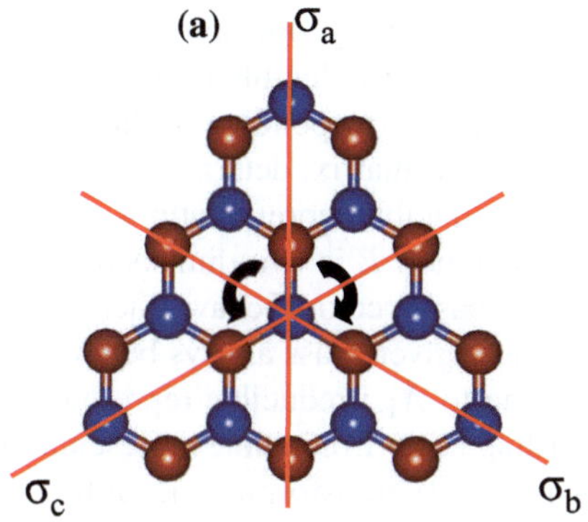

(b)

	E	$2C_3$	$3\sigma_v$
A_1	1	1	1
A_2	1	1	-1
E	2	-1	0

Fig. 4.22 **a** TGQD with all symmetry operations in a C_{3v} symmetry group. Three *red lines* correspond to three reflection planes and two *black arrows* correspond to two rotations over $2\pi/3$. **b** Character table of the C_{3v} symmetry group

symmetry group. TGQDs are structures with a well-defined triangular symmetry. They transform according to symmetry operations of an equilateral triangle, which correspond to the C_{3v} symmetry group. There are six symmetry operations in the group, shown in Fig. 4.22a: identity $\hat{E}$, three reflections $\hat{\sigma_a}$, $\hat{\sigma_b}$, $\hat{\sigma_c}$ with respect to planes going along secants of three triangle's angles, and two rotations $\hat{C}_{1,2}$ over the angle $\pm 2\pi/3$ with a rotational axis going through the center of the triangle. In the Hilbert space, symmetry operators can be represented by unitary matrices [45]. These matrices are matrix representations of operators. All these matrices commute with the TB Hamiltonian matrix: $[H_{TB}, \sigma_\nu] = [H_{TB}, C_i] = 0$, with $\nu = a, b, c$ and $i = 1, 2$, where we used the same notation for operators in a matrix representation as for symmetry operations. Thus, it is possible to classify the energy states according to eigenvalues of the symmetry operators. For example, the matrix corresponding to the reflection operator can have two eigenvalues, $+1$ and -1. One can find eigenstates of the TB Hamiltonian which change (an antisymmetric state) or do not change (a symmetric state) a sign under a reflection with respect to one of the three reflection planes. We want to classify states not with respect to a single operator but with respect to all symmetry operators in a given group. In other words, one has to find a set of basis vectors, which in a simple situation of non-degenerate states (we concentrate on a degeneracy related to the symmetry of the system, not on an accidental degeneracy), do not mix with each other after transformation under all symmetry operations. In this basis, all symmetry operators will be represented by block diagonal matrices. In the case of 1×1 block, after acting on an arbitrary basis vector, there will be no mixing with other basis vectors. In the case of $n \times n$ block, there can be mixing only between n vectors. Such representations are reducible and blocks correspond to the irreducible representations and can not be reduced at the same time for all symmetry operators by any transformation of the basis vectors.

In Fig. 4.22b, we show the character table corresponding to the C_{3v} symmetry group. The left column contains three irreducible representations labeled as

A_1, A_2, E. The top row corresponds to symmetry operators divided into three classes. Elements of the table are characters of irreducible representations, which are traces of matrices in this case. Characters corresponding to the identity operator $\hat{E}$, which is always represented by the unit matrix, determine the dimension of the irreducible representation. Thus, the irreducible representations A_1, A_2 are one-dimensional while the irreducible representation E is two-dimensional. Characters for other symmetry operators describe how basis vectors behave after transformation under symmetry operators. Elements from a given class always behave in the same way. Basis vectors transforming according to A_1 irreducible representation do not change, while these transforming according to A_2 irreducible representation change sign under three reflections. Thus, basis vectors transforming according to A_1 irreducible representation are fully symmetric while these transforming according to A_2 irreducible representation are fully antisymmetric, which is schematically shown in Fig. 4.23a, b, respectively. In the case of the 2D irreducible representation E the situation is more complicated because different linear combinations of two basis vectors can be chosen. One of the choices is such that one basis vector changes sign, and the second one does not, under reflection giving the character (trace) of the representation matrix equal to zero in agreement with the character table. On the other hand, one can choose two basis vectors of the irreducible representation E such that they acquire extra phase $e^{2\pi i/3}$ under rotations, schematically shown in Fig. 4.23c.

We estimate the number of basis vectors a_Γ transforming according to each irreducible representation using [45]

$$a_\Gamma = \frac{1}{h} \sum_i \chi(\hat{R}_i)\chi_\Gamma(\hat{R}_i), \tag{4.46}$$

where $\Gamma = A_1, A_2, E$, $h = 6$ is the number of elements in the group, $\chi(\hat{R}_i)$ and $\chi_\Gamma(\hat{R}_i)$ are characters of the symmetry operator $\hat{R}_i$ of reducible and irreducible representations, respectively. Characters of the reducible representation can be easily evaluated: it is the number of orbitals which remain unchanged under a given symmetry operation. For example, for a triangle from Fig. 4.22a, $\chi(C_{1,2}) = 1$ and $\chi(\sigma_{a,b,c}) = 4$, and $\chi(E) = 22$ as is the number of atoms. Thus, using (4.46) we get $a_{A_1} = 6$, $a_{A_2} = 2$, $a_E = 7$. We can now construct basis vectors for each irreducible representation [45]:

$$\Psi_n^\Gamma = \sum_i D_\Gamma(\hat{R}_i)\hat{R}_i\phi_j, \tag{4.47}$$

where $D_\Gamma(\hat{R}_i)$ is the matrix of an operator $\hat{R}_i$ for Γ irreducible representation. Index j runs over all 22 atomic orbitals but, e.g. for A_1 subspace, only $a_{A_1} = 6$ linearly independent vectors will be obtained, thus $n = 1, 2, \ldots, 6$. We use indices of p_z orbitals from Fig. 4.24. We apply (4.47) first to ϕ_0 and ϕ_1 which for the A_1 representation gives

$$\Psi_1^{A_1} = 1 \cdot \hat{E}\phi_0 + 1 \cdot \hat{\sigma_a}\phi_0 + 1 \cdot \hat{\sigma_b}\phi_0 + 1 \cdot \hat{\sigma_c}\phi_0 + 1 \cdot \hat{C_1}\phi_0 + 1 \cdot \hat{C_2}\phi_0 = 6\phi_0$$

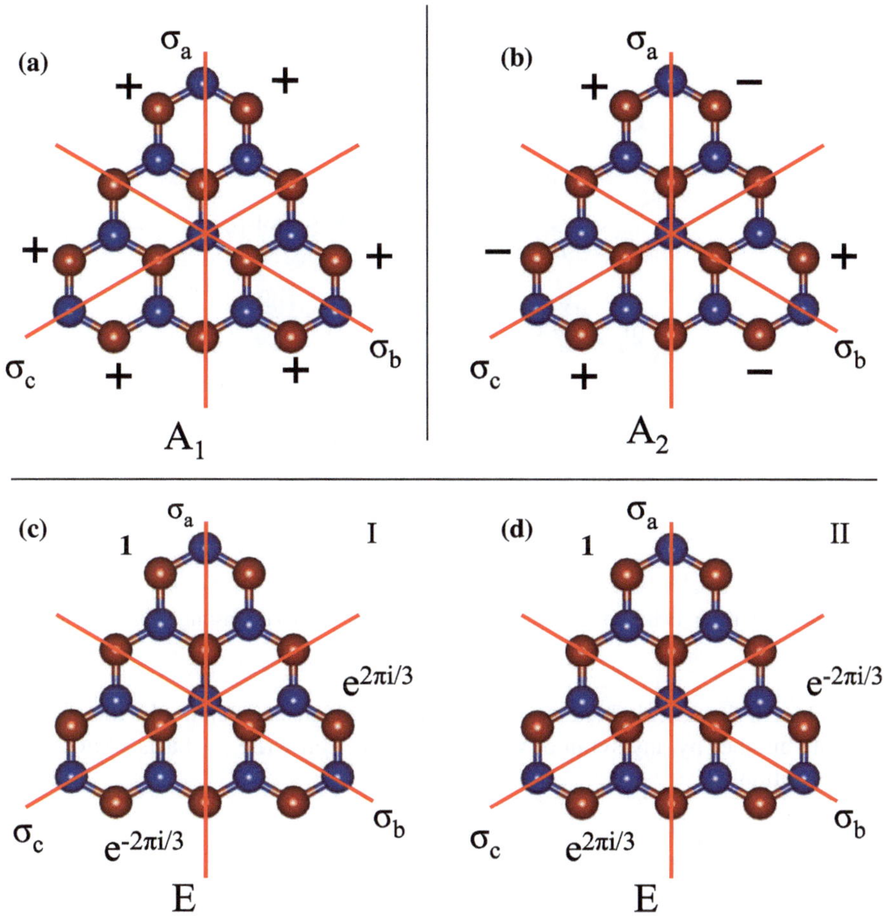

Fig. 4.23 Basis vectors constructed as linear combinations of p_z orbitals of TGQD can be classified according to irreducible representations of the symmetry group. **a** Vectors transforming according to A_1 irreducible representation do not change sign under three reflections (fully symmetric states). **b** Vectors transforming according to A_2 irreducible representation change sign under three reflections (fully antisymmetric states). **c** Vectors transforming according to E irreducible representation acquire extra phase $e^{\pm 2\pi i/3}$ under rotations

$$\begin{aligned} \Psi_2^{A_1} &= 1 \cdot \hat{E}\phi_1 + 1 \cdot \hat{\sigma_a}\phi_1 + 1 \cdot \hat{\sigma_b}\phi_1 + 1 \cdot \hat{\sigma_c}\phi_1 + 1 \cdot \hat{C}_1\phi_1 + 1 \cdot \hat{C}_2\phi_1 \\ &= \phi_1 + \phi_8 + \phi_1 + \phi_{15} + \phi_8 + \phi_{15} = 2(\phi_1 + \phi_8 + \phi_{15}), \end{aligned} \tag{4.48}$$

where in (4.47) $D_{A_1}(\hat{R}_i) = 1$ for all symmetry operators according to the character table shown in Fig. 4.22b. With the help of Fig. 4.24, it is easy to see that $\Psi_2^{A_1}$ can be also obtained by starting from ϕ_8 or ϕ_{15} orbitals. From this we can conclude that all A_1 basis vectors can be obtained using (4.47), starting from ϕ_j for $j = 0, 2, \ldots, 5$, see

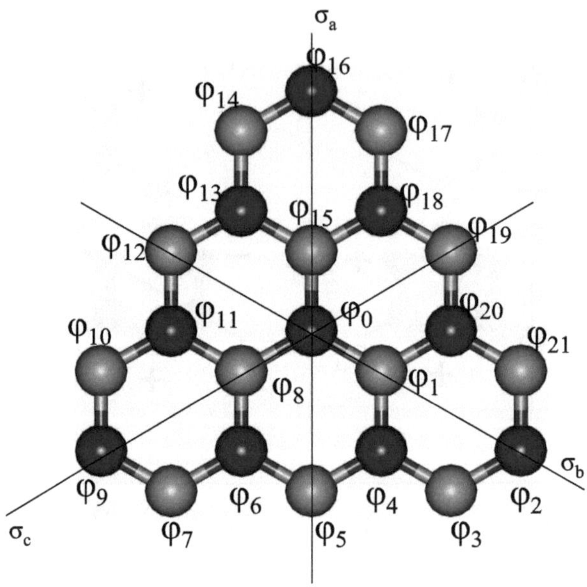

Fig. 4.24 Linking up indices j to all atomic p_z orbitals for TGQD consisting of $N = 22$ atoms

Fig. 4.24. All these orbitals lie in one part of the triangle and can not be transformed one into another by any symmetry operations. We can write A_1 basis vectors after normalization as

$$\begin{aligned}
\Psi_1^{A_1} &= \phi_0 \\
\Psi_2^{A_1} &= \frac{1}{\sqrt{3}}(\phi_1 + \phi_8 + \phi_{15}) \\
\Psi_3^{A_1} &= \frac{1}{\sqrt{3}}(\phi_2 + \phi_9 + \phi_{16}) \\
\Psi_4^{A_1} &= \frac{1}{\sqrt{6}}(\phi_3 + \phi_{21} + \phi_{14} + \phi_7 + \phi_{10} + \phi_{17}) \\
\Psi_5^{A_1} &= \frac{1}{\sqrt{6}}(\phi_4 + \phi_{20} + \phi_{13} + \phi_6 + \phi_{11} + \phi_{18}) \\
\Psi_6^{A_1} &= \frac{1}{\sqrt{3}}(\phi_5 + \phi_{19} + \phi_{12}) .
\end{aligned}$$

These states are fully symmetric, which was schematically shown in Fig. 4.23a. In a similar way, one can construct basis vectors transforming according to the irreducible representation A_2. We apply (4.47) first, e.g. to ϕ_2 and ϕ_4, getting

$$\Psi_1^{A_2} = 1 \cdot \hat{E}\phi_2 - 1 \cdot \hat{\sigma_a}\phi_2 - 1 \cdot \hat{\sigma_b}\phi_2 - 1 \cdot \hat{\sigma_c}\phi_2 + 1 \cdot \hat{C}_1\phi_2 + 1 \cdot \hat{C}_2\phi_2$$

$$= \phi_2 - \phi_9 - \phi_2 - \phi_{16} + \phi_9 + \phi_{16} = 0$$

$$\Psi_2^{A2} = 1 \cdot \hat{E}\phi_4 - 1 \cdot \hat{\sigma_a}\phi_4 - 1 \cdot \hat{\sigma_b}\phi_4 - 1 \cdot \hat{\sigma_c}\phi_4 + 1 \cdot \hat{C}_1\phi_4 + 1 \cdot \hat{C}_2\phi_4$$
$$= \phi_4 - \phi_{20} - \phi_{13} - \phi_6 + \phi_{11} + \phi_{18}, \tag{4.49}$$

where $D_{A_2}(\sigma_v) = -1$ and $D_{A_2}(C_i) = 1$ according to the character table shown in Fig. 4.22b. The first vector vanishes identically. This gives a clue that the starting orbital can not lie on a line associated with one of reflection planes. We have only $a_{A_2} = 2$ basis vectors, and the second one can be obtained starting from ϕ_3. We can write A_2 basis vectors after normalization as

$$\Psi_1^{A_2} = \frac{1}{\sqrt{6}} (\phi_3 - \phi_{21} - \phi_{14} - \phi_7 + \phi_{10} + \phi_{17})$$
$$\Psi_2^{A_2} = \frac{1}{\sqrt{6}} (\phi_4 - \phi_{20} - \phi_{13} - \phi_6 + \phi_{11} + \phi_{18}) .$$

These states are fully antisymmetric which was schematically shown in Fig. 4.23b.

We construct basis vectors transforming according to the irreducible representation E. In order to do this, we define irreducible representations for symmetry operators because only characters of these matrices are known, see Fig. 4.22b. We chose the following unitary matrices,

$$D_E(\hat{E}) = \begin{pmatrix} 1 & 0 \\ 0 & 1 \end{pmatrix}, D_E(\hat{\sigma_a}) = \begin{pmatrix} 0 & 1 \\ 1 & 0 \end{pmatrix},$$
$$D_E(\hat{\sigma_b}) = \begin{pmatrix} 0 & e^{-2i\pi/3} \\ e^{2i\pi/3} & 0 \end{pmatrix}, D_E(\hat{\sigma_c}) = \begin{pmatrix} 0 & e^{2i\pi/3} \\ e^{-2i\pi/3} & 0 \end{pmatrix},$$
$$D_E(\hat{C}_1) = \begin{pmatrix} e^{2i\pi/3} & 0 \\ 0 & e^{-2i\pi/3} \end{pmatrix}, D_E(\hat{C}_2) = \begin{pmatrix} e^{-2i\pi/3} & 0 \\ 0 & e^{2i\pi/3} \end{pmatrix}. \tag{4.50}$$

We apply (4.47) first to ϕ_1. We have four matrix elements in each matrix, (4.50), so we obtain four functions

$${}^{11}\Psi_1^E = {}^{11}D_E(\hat{E})\hat{E}\phi_1 + {}^{11}D_E(\hat{\sigma_a})\hat{\sigma_a}\phi_1 + {}^{11}D_E(\hat{\sigma_b})\hat{\sigma_b}\phi_1$$
$$+ {}^{11}D_E(\hat{\sigma_c})\hat{\sigma_c}\phi_1 + {}^{11}D_E(\hat{C}_1)\hat{C}_1\phi_1 + {}^{11}D_E(\hat{C}_2)\hat{C}_2\phi_1$$
$$= 1 \cdot \phi_1 + 0 \cdot \phi_1 + 0 \cdot \phi_{15} + 0 \cdot \phi_8 + e^{-2i\pi/3} \cdot \phi_8 + e^{2i\pi/3} \cdot \phi_{15}$$
$$= \phi_1 + e^{-2i\pi/3}\phi_8 + e^{2i\pi/3}\phi_{15}$$
$${}^{12}\Psi_1^E = e^{-2i\pi/3}\left(\phi_1 + e^{-2i\pi/3}\phi_8 + e^{2i\pi/3}\phi_{15}\right)$$
$${}^{21}\Psi_1^E = e^{2i\pi/3}\left(\phi_1 + e^{2i\pi/3}\phi_8 + e^{-2i\pi/3}\phi_{15}\right)$$
$${}^{22}\Psi_1^E = \phi_1 + e^{2i\pi/3}\phi_8 + e^{-2i\pi/3}\phi_{15}.$$

It is clearly seen that ${}^{11}\Psi_1^E$ and ${}^{12}\Psi_1^E$ are linearly dependent. Similarly ${}^{21}\Psi_1^E$ and ${}^{22}\Psi_1^E$ are linearly dependent. Thus, two linearly independent basis vectors can be chosen as

$$\begin{aligned}\Psi_{11}^E &= {}^{11}\Psi_1^E = \phi_1 + e^{-2i\pi/3}\phi_8 + e^{2i\pi/3}\phi_{15} \\ \Psi_{12}^E &= {}^{22}\Psi_1^E = \phi_1 + e^{2i\pi/3}\phi_8 + e^{-2i\pi/3}\phi_{15}.\end{aligned}$$

We see that orbitals in these vectors are obtained by starting with one of them and rotating it over $\pm 2\pi/3$. Thus, all E basis vectors can be found starting from orbitals ϕ_j for $j = 1, 2, \ldots, 7$, which lie in $1/3$ of the triangle and can not be transformed one into another by any of the two rotations, see Fig. 4.24. These vectors with appropriate normalization, with the help of Fig. 4.24, can be shortly written as

$$\begin{aligned}\Psi_{j1}^E &= \frac{1}{\sqrt{3}}\left(\phi_j + e^{-2i\pi/3}\phi_{j+7} + e^{2i\pi/3}\phi_{j+14}\right) \\ \Psi_{j2}^E &= \frac{1}{\sqrt{3}}\left(\phi_j + e^{2i\pi/3}\phi_{j+7} + e^{-2i\pi/3}\phi_{j+14}\right),\end{aligned}$$

for $j = 1, 2, \ldots, 7$. Having all basis vectors, the TB has a block diagonal form, shown in Fig. 4.25. Three blocks corresponding to each irreducible representation are visible. Matrix elements between basis vectors transforming according to different irreducible representations vanish identically. For example, for $\langle \Psi_3^{A_1} | H_{TB} | \Psi_{11}^E \rangle$ one gets

$$\begin{aligned}\langle \Psi_3^{A_1} | H_{TB} | \Psi_{11}^E \rangle = \frac{1}{3}(&\langle \phi_2 | H_{TB} | \phi_1 \rangle + e^{-2i\pi/3}\langle \phi_2 | H_{TB} | \phi_8 \rangle + e^{2i\pi/3}\langle \phi_2 | H_{TB} | \phi_{15} \rangle \\ &+ \langle \phi_9 | H_{TB} | \phi_1 \rangle + e^{-2i\pi/3}\langle \phi_9 | H_{TB} | \phi_8 \rangle + e^{2i\pi/3}\langle \phi_9 | H_{TB} | \phi_{15} \rangle \\ &+ \langle \phi_{16} | H_{TB} | \phi_1 \rangle + e^{-2i\pi/3}\langle \phi_{16} | H_{TB} | \phi_8 \rangle + e^{2i\pi/3}\langle \phi_{16} | H_{TB} | \phi_{15} \rangle),\end{aligned}$$

where due to the symmetry of the system

$$\begin{aligned}\langle \phi_2 | H_{TB} | \phi_1 \rangle &= \langle \phi_9 | H_{TB} | \phi_8 \rangle = \langle \phi_{16} | H_{TB} | \phi_{15} \rangle, \\ \langle \phi_2 | H_{TB} | \phi_8 \rangle &= \langle \phi_9 | H_{TB} | \phi_{15} \rangle = \langle \phi_{16} | H_{TB} | \phi_8 \rangle, \\ \langle \phi_2 | H_{TB} | \phi_{15} \rangle &= \langle \phi_9 | H_{TB} | \phi_1 \rangle = \langle \phi_{16} | H_{TB} | \phi_1 \rangle,\end{aligned} \tag{4.51}$$

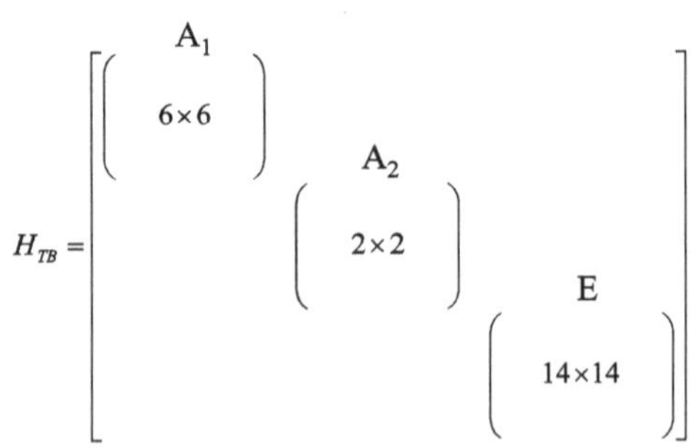

Fig. 4.25 The scheme of a TB Hamiltonian matrix written in a basis of vectors transforming according to irreducible representation of an equilateral triangle. The matrix takes a block diagonal form

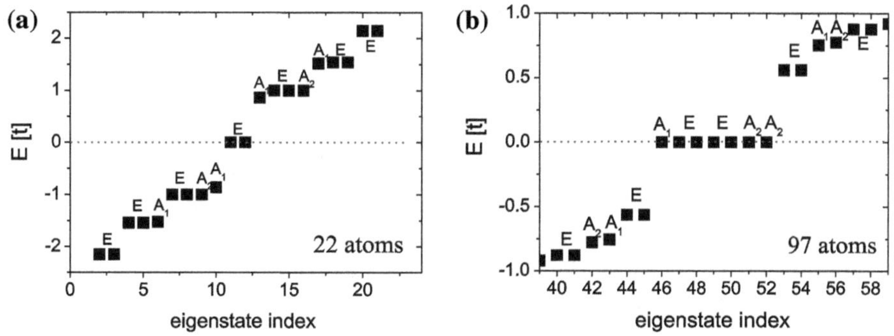

Fig. 4.26 Energy spectra of TB Hamiltonian for TGQDs with **a** $N = 22$ and **b** $N = 97$ atoms. Each energy level transforms according to the irreducible representation of C_{3v} symmetry group

which was obtained with the help of Fig. 4.24. Finally, one gets

$$\begin{aligned}\langle \Psi_3^{A_1}|H_{TB}|\Psi_{11}^{E}\rangle = \frac{1}{3}(\langle\phi_2|H_{TB}|\phi_1\rangle \left(1 + e^{-2i\pi/3} + e^{2i\pi/3}\right) \\ + \langle\phi_2|H_{TB}|\phi_8\rangle \left(1 + e^{-2i\pi/3} + e^{2i\pi/3}\right) \\ + \langle\phi_2|H_{TB}|\phi_15\rangle \left(1 + e^{-2i\pi/3} + e^{2i\pi/3}\right) = 0,\end{aligned}$$

for arbitrary matrix elements because $1 + e^{-2i\pi/3} + e^{2i\pi/3} = 0$.

In Fig. 4.26a we show the energy spectrum of a TGQD from Fig. 4.22a with each energy level classified by the symmetries of the corresponding eigenstates. There are $N_{deg} = 2$ degenerate zero-energy states. They transform according to the E irreducible representation. The highest (lowest) state of the valence (conduction) band transforms according to the A_1 irreducible representation. In Fig. 4.26b, we show the energy spectrum of the structure consisting of $N = 97$ atoms with $N_{deg} = 7$ degenerate zero-energy states. Here, zero-energy states are characterized by different symmetries. There are two states transforming according to A_2 and E, and one transforming according to A_1 irreducible representation. Thus, it is clearly seen that the zero-energy degeneracy is not related to the symmetry of the system. Such degeneracy must therefore be related to a "hidden symmetry". We note that for all studied structures the number of states with a given symmetry in the degenerate shell, N_{deg}^{Γ}, can be evaluated using the following expressions, for n-integer,

$$N_{deg}^{A_1} = \begin{cases} \text{int}\left(\frac{\text{N}_{\text{deg}}+2}{6}\right) & N_{deg} \neq 6n - 1 \\ \text{int}\left(\frac{\text{N}_{\text{deg}}-4}{6}\right) & N_{deg} = 6n - 1 \end{cases}$$

$$N_{deg}^{A_2} = \begin{cases} \text{int}\left(\frac{N_{deg}+5}{6}\right) & N_{deg} \neq 6n-4 \\ \text{int}\left(\frac{N_{deg}}{6}\right) & N_{deg} = 6n-4 \end{cases}$$

$$N_{deg}^{E} = \left\{ \text{int}\left(\frac{N_{deg}+1}{3}\right) \quad N_{deg} = 1, 2, ... \right.$$

Additionally, in the energy spectra shown in Fig. 4.26a, b, the highest (lowest) state of the valence (conduction) band transforms according to A_1 and E irreducible representation, respectively. We note that for all studied structures the symmetry of these states confirms the following

$$\begin{array}{ccc} A_1 & \text{for} & N_{edge} = 3n-1 \\ E & \text{for} & \begin{array}{c} N_{edge} = 3n \\ N_{edge} = 3n-2 \end{array} \end{array}.$$

for n-integer. The symmetry classification of zero-energy states is relevant to the Gramm-Schmit orthogonalization of linearly independent N_{deg} vectors obtained in the previous section.

4.3.5 The Effect of Spin-Orbit Coupling

In Sect. 4.2.3 we have shown that the spin-orbit coupling induces a topological spin Hall effect at the edges of graphene zigzag nanoribbons. We will now study the effect of the spin-orbit coupling at the edges of triangular zigzag quantum dots. This can be done by either diagonalizing the four-band tight-binding Hamiltonian given by (4.19), or the effective Kane-Mele Hamiltonian given by (4.24), also allowing us to test the validity of Kane-Mele approximation.

The effect of the spin-orbit coupling on the zero-energy states of a TGQD as obtained by the four-band tight-binding and Kane-Mele Hamiltonians are shown in Fig. 4.27 for a triangular quantum dot with $N = 97$ atoms. As discussed in the previous section, there are $N_{deg} = 7$ zero-energy states for spin-up and for spin-down electrons in the absence of spin-orbit coupling. As we turn the spin-orbit coupling on, spin and orbital degeneracy is lifted. Note that the four-orbital tight-binding method mixes up and down spin states and S_z is not a good quantum number anymore. This is why in Fig. 4.27 we plot 14 states including the spin degree of freedom, instead of 7. However, for parameters corresponding to carbon atoms, the spin contamination is very small, less than 0.1 %. This justifies the use of the Kane-Mele Hamiltonian conserving S_z. There are other small differences between the four-orbital tight-binding and Kane-Mele approximation calculations: first, Kane-Mele Hamiltonian conserves the electron-hole symmetry, thus the spectrum is symmetric around the Fermi level, in contrast with the four-orbital results. Moreover, the dependence ofenergy levels

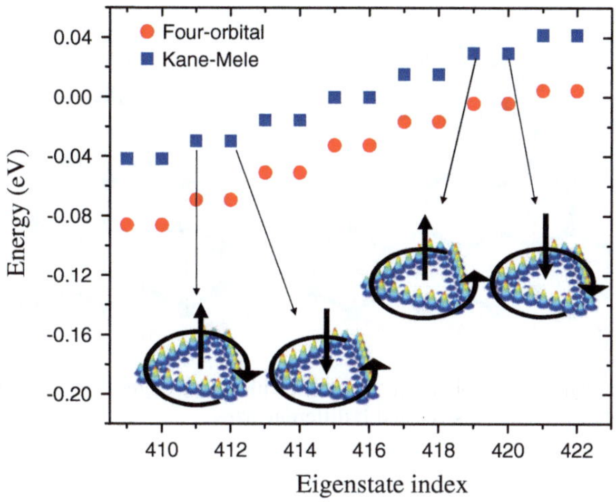

Fig. 4.27 A comparison between the effect of spin-orbit coupling on the zero-energy states of a TGQD with $N = 97$ atoms and $N_{deg} = 7$ zero-energy states calculated using four-band tight-binding, (4.19), and Kane-Mele Hamiltonian, (4.24)

on the eigenstate index is linear for zero-energy states in the Kane-Mele model, whereas a small non-linearity is detected in the four-orbital spectrum. The wave-functions obtained using the two methods are, however, very similar.

In the inset of Fig. 4.27 we show the probability densities corresponding to two pairs of states. Each pair is connected by Kramers degeneracy, thus up and down spins couple to angular momentum states rotating in opposite directions. The spin-orbitals of the second pair, shown on the right hand side of the figure, share the same orbitals as the first pair but are coupled to opposite spins. These results show that, if a new electron is added to the system, it will rotate in a direction dictated by its spin. Such structure can be used as a single-spin filter device. Note that although the physics described here is similar to the spin-orbit coupling in graphene ribbons, there is an important difference: the triangular graphene dot has only one type of edge, i.e. involves only one sublattice, whereas the opposite edges of the graphene ribbon lie on two different sublattices. This has important implications if electron-electron interactions are taken into account: for ribbons the opposite edges interact antiferromagnetically [46–48], while triangular dot edges are ferromagnetically coupled as we will see in Chap. 6.

4.4 Bilayer Triangular Graphene Quantum Dots with Zigzag Edges

In Sect. 2.4.4 we showed that it is possible to open a gap and control the energy spectrum of a bilayer graphene by an external electric field which creates a potential difference between the two layers. In a bilayer triangular graphene quantum dot, the

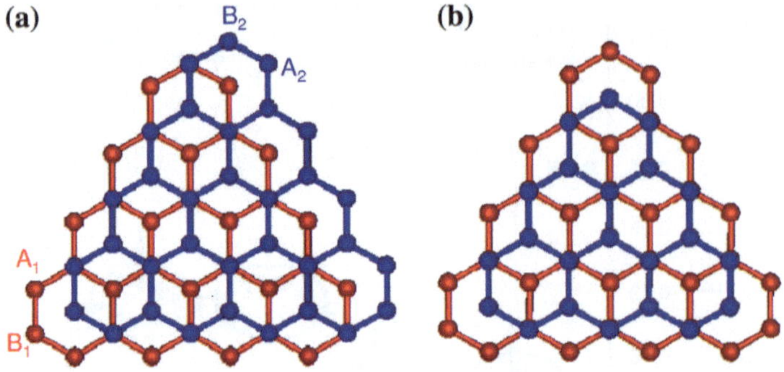

Fig. 4.28 Bilayer triangular graphene quantum dot with zigzag edges, constructed using two single layer quantum dots with **a** equal sizes and **b** different sizes

physics becomes even more interesting due to the presence of zero-energy states. In this section we will investigate the effect of inter-layer coupling and an external perpendicular electric field on the zero-energy states of bilayer triangular quantum dots with zigzag edges (BTGQD).

Figure 4.28 shows two possibilities for building a bilayer triangular graphene quantum dot with zigzag edges (BTGQD) using two single-layer triangular quantum dots (TGQD) of comparable sizes. As in Sect. 2.4.4, we consider AB Bernal stacking, where the A sublattice of the top-layer (A2, shown in blue color) is on top of the B sublattice of the bottom-layer (B1, shown in red). In Fig. 4.28a, the two TGQDs are of the same size. In this configuration, however, on one edge of the triangle not all the A2 atoms have a B1 partner as required by Bernal stacking. A more natural configuration choice is shown in Fig. 4.28b. The top-layer triangle has its floating atoms removed, making it smaller than the bottom layer triangle. Such a bilayer construction has the interesting property of having an odd number of degenerate states as we will discuss in the following.

In order to study single particle properties, we diagonalize the tight-binding Hamiltonian given by

$$H_{TB} = \sum_{ij\sigma} t_{ij} c_{i\sigma}^{\dagger} c_{j\sigma} + \sum_{i\sigma} V_i c_{i\sigma}^{\dagger} c_{i\sigma}, \tag{4.52}$$

where we now need to include the inter-layer coupling. The tight-binding parameters t_{ij} are fixed to $t = -2.8\,\mathrm{eV}$ for in-plane nearest neighbors i and j and $t_{\perp} = 0.4\,\mathrm{eV}$ for inter-layer hopping between A2 and B1 atoms. The effect of the potential difference induced by an external perpendicular electric-field **E** is taken into account through $V_i = -\Delta V/2$ for the bottom-layer atoms and $V_i = \Delta V/2$ for the top-layer atoms. Figure 4.29 shows the energy spectrum near the Fermi level for $\Delta V = 0$ for a BTGQD consisting of 622 atoms in the bottom and 573 atoms in the top-layer. If we take $t_{\perp} = 0$, the two triangles are decoupled and we find 22 zero-energy states in the bottom-layer and 21 zero-energy states in the top-layer, for a total of 43 zero-energy

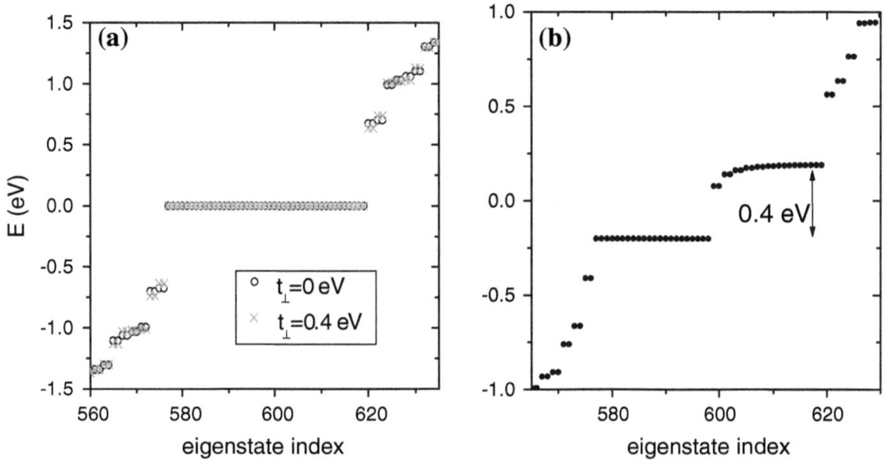

Fig. 4.29 Single-particle tight-binding spectrum. **a** The bilayer quantum dot consisting of 1,195 atoms has 43 zero-energy states. **b** When an electric-field is applied, the degeneracy between the 21 *top-layer* zero-energy states and 22 *bottom-layer* zero-energy states is lifted. Reprinted from [49]

states, consistent with previous work on single-layer TGQDs [43, 50–56]. Turning on $t_{\perp}$ to 0.4 eV does not affect the zero-energy states. The effect of applying an electric field, e.g. $\Delta V = 0.4$ eV, is shown in Fig. 4.29b. The energy of the 21 zero-energy states corresponding to the top-layer is pushed up by 0.4 eV with respect to the bottom-layer zero-energy states. Note that the bottom-layer zero-energy states do not experience any dispersion unlike the top-layer zero-energy states. This is due to the fact that they lie strictly on A1 sites which are not coupled to the top layer, whereas the top layer zero-energy states, which lie on B2, do couple to the bottom layer. The ability of controlling the relative position of zero-energy states presents an interesting opportunity to control the charge and spin of the zero-energy states. We will discuss electron-electron interactions and magnetic properties of bilayer quantum dots in detail in Chap. 6.

4.5 Triangular Mesoscopic Quantum Rings with Zigzag Edges

In previous chapter we discussed the manipulation of the zero-energy shell by stacking two TGQDs vertically. Here we discuss the effect of increasing the role of the zigzag edge by creating zigzag edge holes in TGQDs [57]. TGQRs can be fabricated using carbon nanotubes (CNT) as a mask in the etching process. One can place a CNT over the graphene sheet along a given crystallographic direction and cover atoms lying below, e.g., along a zigzag direction. Three carbon nanotubes can be arranged in a triangular shape, along three zigzag edges, shown on the left in Fig. 4.30. As a result one expects to obtain a triangular structure with well-defined zigzag edges and a hole in the center, as shown on the right in Fig. 4.30. The full

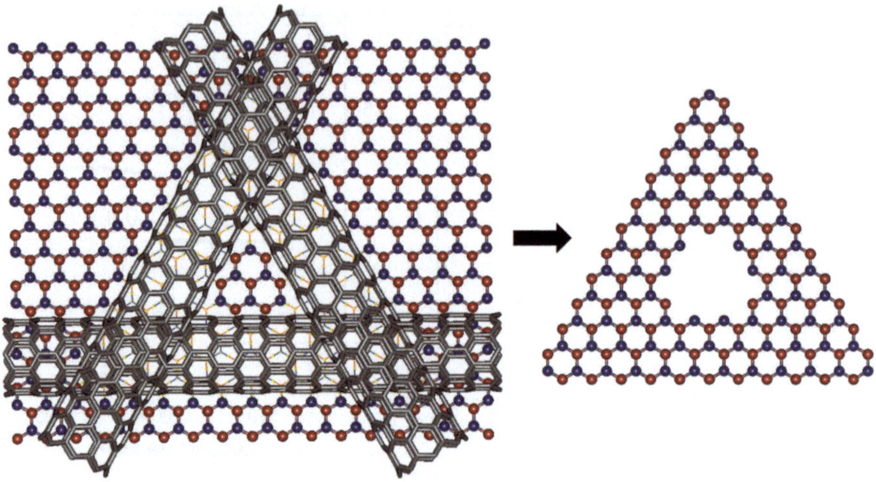

Fig. 4.30 Proposed experimental method for designing TGQR. Three CNTs arranged in *equilateral triangle* along zigzag edges play the role of a mask. By using etching methods one can obtain TGQR with well defined edges. The circumference of CNT determines the width of TGQR. *Red* and *blue* colors distinguish between two sublattices in the honeycomb graphene lattice. Reprinted from [57]

TGQD consists of $N_{out}^2 + 4N_{out} + 1$ atoms, where $N_{out} = N_{edge}$. The small removed triangle consists of $N_{inn}^2 + 4N_{inn} + 1$ atoms, where N_{inn} is the number of edge atoms on one inner edge. The resulting TGQR has $N = N_{out}^2 - N_{inn}^2 + 4(N_{out} - N_{inn})$ atoms. Its width satisfies $N_{out} - N_{inn} = 3(N_{width} + 1)$, where N_{width} is the width counted in the number of benzene rings. The structure shown on the right of Fig. 4.30 has $N_{width} = 2$. We note that while outer edges are built of atoms of type A, inner edges are built of atoms of type B.

4.5.1 Energy Spectrum

In the full triangle, the imbalance between the number of A type (N_A) and B type (N_B) of atoms in the bipartite honeycomb graphene lattice, proportional to N_{edge}, leads to the appearance of zero-energy states in the TB model in the nearest-neighbor approximation. The number of zero-energy states is $N_{deg} = |N_A - N_B|$, as shown in the subsection 4.3.2. Removing a small triangle from the center lowers the imbalance between two types of atoms in the structure, leading to a decreased number of zero-energy states. The degeneracy of the zero-energy shell in TGQRs can be defined as $N_{deg} = |N_A - N_B| = 3(N_{width} + 1)$. Thus, the number of zero-energy states in TGQRs only depends on the width of the ring, and not on its size.

In Fig. 4.31, we show the single-particle spectra for TGQRs obtained by diagonalizing the TB Hamiltonian, (4.1). Figure 4.31a shows the energy spectrum for TGQR with $N_{width} = 2$ consisting of $N = 171$ atoms and shown in Fig. 4.30 on

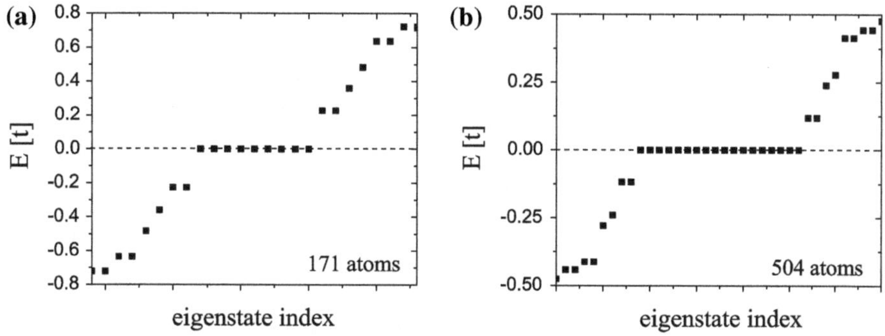

Fig. 4.31 Single particle TB levels for TGQR with **a** $N_{width} = 2$, consisting of 171 atoms and **b** $N_{width} = 5$, consisting of 504 atoms. The degeneracy at the Fermi level (*dashed line*) is a function of the width $N_{deg} = 3(N_{width} + 1)$, for (**a**) $N_{deg} = 9$ and for (**b**) $N_{deg} = 18$. Reprinted from [57]

the right. It has $N_{out} = 11$ and $N_{inn} = 2$ and the number of zero-energy states is $N_{deg} = 9$. Figure 4.31b shows TB spectrum of a TGQR with $N_{width} = 5$, consisting of 504 atoms. It has $N_{out} = 21$ and $N_{inn} = 3$, giving $N_{deg} = 18$, consistent with our formula $N_{deg} = 3(N_{width} + 1)$. We note that the states of the zero-energy shell consist of orbitals belonging to one type of atoms indicated with the red color in Fig. 4.30, and lie mostly on the outer edge. On the other hand, the other states close to the Fermi level consist of orbitals belonging to both sublattices, but lie mostly on the inner edge (not shown here). This fact has implications for the magnetic properties of the system, described in the Sect. 4.2.

4.6 Hexagonal Mesoscopic Quantum Rings

In order to investigate the dependence of the electronic properties of graphene quantum rings on ring geometry we consider here hexagonal mesoscopic quantum rings [58]. Below, we present a method of constructing hexagonal mesoscopic quantum rings. We first consider six independent nanoribbons, then bring them together by turning on the hopping between the connecting atoms. In Fig. 4.32 we show two sets of six graphene ribbons arranged in a hexagonal ring. On the left side, we show the thinnest possible ribbons with one benzene ring width, denoted as $W = 1$. Each of them consists of 16 atoms. The length, $L = 4$, is measured by the number of one type of atoms in the upper row, so the final ring is built out of $N = 96$ atoms. Small black arrows in the bottom enlargement indicate bonds and hopping integrals between nearest neighbors in the TB model between neighboring ribbons, two arrows in the case of the thinnest structures. The number of such connecting atoms increases with increasing width as seen on the right hand side of Fig. 4.32. The thicker ribbon, $W = 2$, has its length identical to that from the left side $L = 4$. In this case there are three connecting atoms. Three small black arrows in the bottom enlargement indicate three bonds. The final ring is built out of $N = 126$ atoms. By connecting the

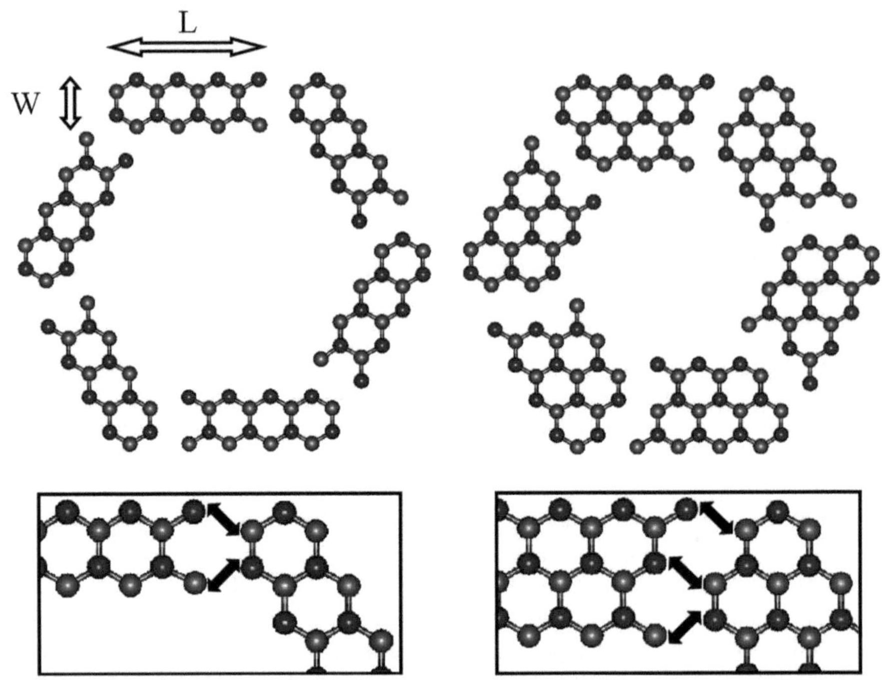

Fig. 4.32 Construction of ring structures from six ribbon-like units. On the *left*, there are six thinnest possible ribbons (one benzene ring thick denoted as $W = 1$) arranged in a hexagonal ring structure. The length of each ribbon is given by $L = 4$, the number of one type of atoms in one row. Each ribbon consists of 16 atoms which gives a total of $N = 96$ atoms in a ring. On the *right*, there are six ribbons with width $W = 2$ (two benzene ring thick). Each of them consists of 21 atoms giving a total of $N = 126$ atoms in a ring. We create a thicker ring with a similar length $L = 4$ but a smaller antidot inside. *Small black arrows* in the *bottom* enlargement indicate bonds and hopping integrals between nearest neighbors in the TB model between neighboring ribbons. Reprinted from [58]

neighboring ribbons with different lengths and widths, we create rings with different single-particle spectra.

4.6.1 Energy Spectrum

In Fig. 4.33 we show the single-particle energy levels near the Fermi level obtained by diagonalizing the TB Hamiltonian, (4.1), for rings with length $L = 8$ and different widths W. The thinnest ring, $W = 1$, consists of $N = 192$ atoms. For this structure we observe nearly degenerate shells of energy levels separated by gaps. Each shell consists of six levels: two single and two doubly degenerate states. The first shell over the Fermi level is almost completely degenerate, while in the second one the degeneracy is slightly removed. We note that for rings with different lengths, the gap

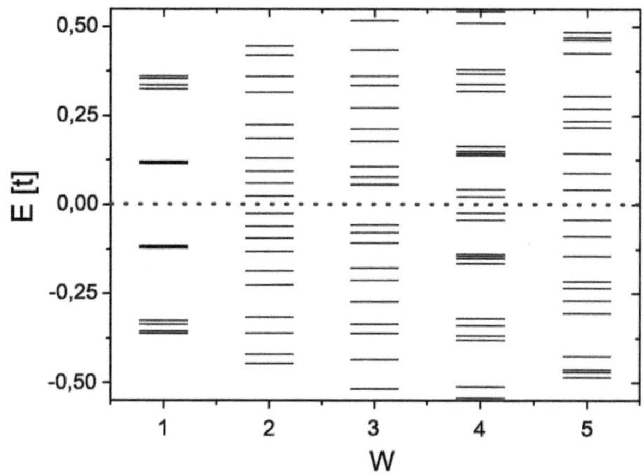

Fig. 4.33 Single particle spectrum near Fermi level for ring structures with $L = 8$, different widths W and $t' = t$ (see Fig. 4.34). The shell structure is clearly observed only for the thinnest ring $W = 1$. *Dotted blue line* indicates the location of Fermi energy. Reprinted from [58]

between the first and second shell is always larger than the gap at the Fermi level. With increasing width of the ring, the spectrum changes completely. For the rings with width $W = 2$ and $N = 270$ atoms, $W = 3$ and $N = 336$ atoms, and $W = 5$ and $N = 432$ atoms, shells are not visible. For $W = 4$ and $N = 390$ atoms we observe the appearance of shells separated by gaps further from the Fermi level but the splitting between levels in these shells is much stronger in comparison to the thinnest ring. We note that for $W \geq 2$, although we do not observe a clear pattern of shells around the Fermi level, single shells of six levels separated from the rest of the spectrum by gaps appear far away from the Fermi energy in some cases.

In order to have a better understanding of the structure of the TB spectra, in Fig. 4.34 we show the evolution of single-particle energies from six independent ribbons to a ring as the hopping t' between the ribbons is increased. To achieve this, we first diagonalize the TB Hamiltonian matrix for a single ribbon. We then take six such ribbons and create the Hamiltonian matrix in the basis of the eigenvectors of six ribbons. Here, the matrix has a diagonal form. All energy levels are at least sixfold degenerate. Next, using the six-ribbon basis, we write hopping integrals corresponding to connecting atoms between neighboring ribbons indicated by small black arrows in Fig. 4.32. By slowly turning on the hopping integrals and diagonalizing the Hamiltonian at every step, we can observe the evolution of the spectrum from single-particle states of six independent ribbons to a ring.

The hopping integrals between connecting atoms of neighboring ribbons are indicated by t' in Fig. 4.34. For the thinnest ring (Fig. 4.34a), each ribbon consists of 32 atoms. There are only two connecting atoms between neighboring ribbons, giving only two hopping integrals t' between each two ribbons in the nearest-neighbor approximation. We see that their influence is very small and sixfold degenerate states evolve into shells with a very small splitting between levels. We note that this

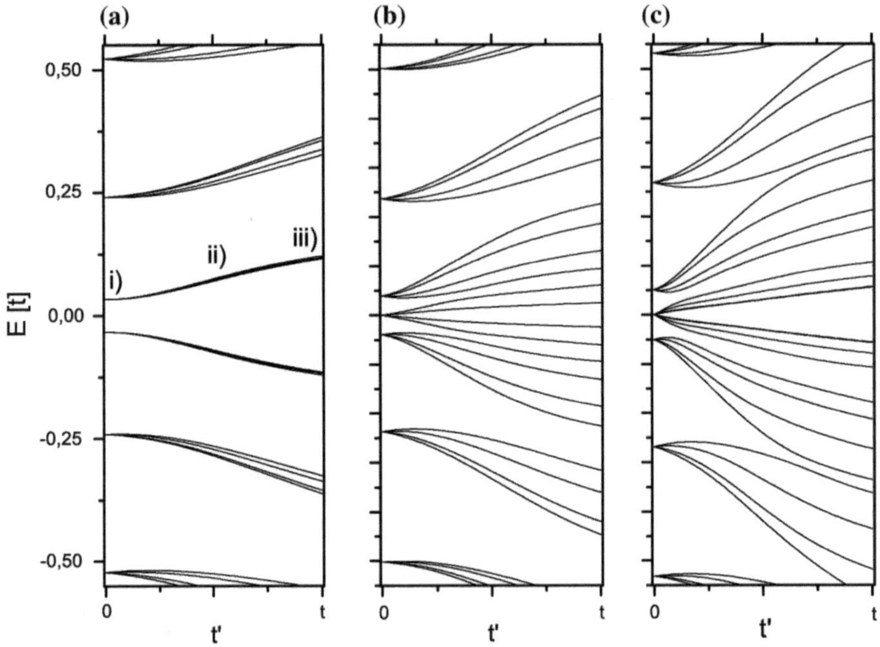

Fig. 4.34 The evolution of the single particle spectrum from six independent ribbons with $L = 8$ to a hexagonal ring structure spectrum. t' indicate hopping integrals between neighboring ribbons. **a** For the *thinnest* ring $W = 1$ six fold degeneracy is slightly removed, preserving a shell structure. For *thicker* structures (**b** and **c**, $W = 2$ and $W = 3$ respectively) the six fold degeneracy is strongly lifted and shell structure is not observed. Reprinted from [58]

splitting is a bit stronger for higher energy levels, but due to large gaps between consecutive levels of a single ribbon, the shell structure is still clearly observed. For the thicker structures (Fig. 4.34b, c), the evolution of the spectrum has a more complicated behavior. For a given ring, each ribbon consists of different number of two types of atoms giving rise to zero-energy edge states [59]. With increasing width, the number of zero-energy states increases as well as the number of connecting atoms and equally the number of t' hopping integrals (see enlargement in Fig. 4.32). This causes a stronger splitting of levels for thicker rings in comparison to the thinnest one. Thus, the spectrum of the thicker ring close to the Fermi level is due to the splitting of zero-energy states of independent ribbons. For $W = 2$ (one zero-energy state) and $W = 3$ (two zero-energy states), each ribbon consists of 45 and 56 atoms respectively, and the evolution of their spectrum is similar. The degeneracy is strongly lifted and no shell structure is observed.

In order to illuminate the influence of t' hopping integrals on the thinnest ring spectrum, in Fig. 4.35, we also show the electronic densities for the first shell over the Fermi level for three different values of t' (indicated in Fig. 4.35a). For $t' = 0$, there are six independent ribbons and the first shell is perfectly sixfold degenerate. The electronic charge density in each ribbon is larger on the two atoms with only

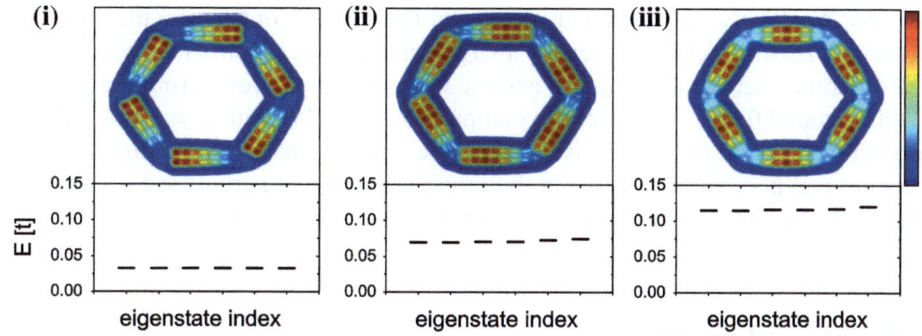

Fig. 4.35 Energy levels and corresponding total electronic densities for the first six states over the Fermi level for the thinnest structure $W = 1$ with $L = 8$ and $N = 192$ atoms, for **a** $t' = 0$, **b** $t' = 0.5t$, **c** $t' = t$. The three values of t' hopping integrals are indicated in Fig. 4.34a. Reprinted from [58]

one bond (see Fig. 4.32) and gradually decreases along the length. For $t' = 0.5t$ the total energy of the shell increases and the degeneracy is slightly removed. Here, the highest peak of the electronic charge density is moved towards the center of each ribbon in comparison to the case of $t' = 0$. Increasing t' to t causes an increase of the total energy of the shell and the highest peak of the electronic charge density is now perfectly in the middle of each arm of the ring. Thus, both the electronic charge density and the energy of levels change slightly during the gradual transition of ribbons into a hexagonal ring structure.

We find degenerate shells near the Fermi energy only for the thinnest rings $W = 1$. In Fig. 4.36 we show the low-energy spectrum for two thinnest rings with different

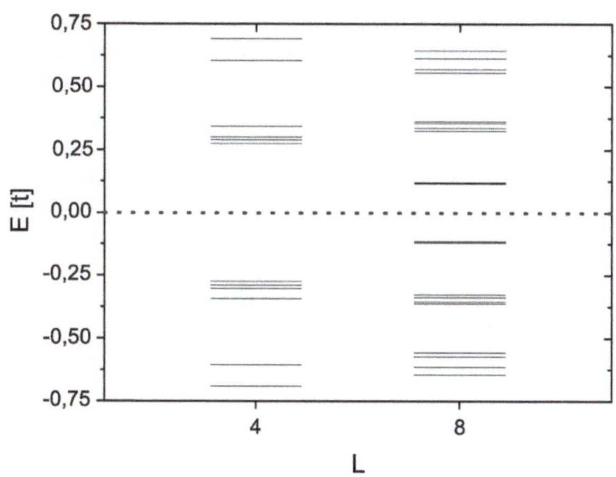

Fig. 4.36 Single particle spectrum near Fermi level for the thinnest ring structures $W = 1$ with length $L = 8$ and $L = 4$. The shell structure is clearly observed. The splitting between levels in the first shell is smaller for larger structure. *Dotted blue line* indicate the location of the Fermi energy. Reprinted from [58]

lengths. We clearly see shells with six levels. The splitting of levels of the first shell above the Fermi level is smaller for a larger ring. For a ring structure with $L = 4$ the difference between the highest and the lowest energy of levels forming the first shell is around $0.069t \simeq 0.17\,\mathrm{eV}$. In comparison, for a ring with $L = 8$ this value is around $0.006t \simeq 0.015\,\mathrm{eV}$. Thus we conclude that for smaller rings single-particle energies can play important role in the properties of many-particle states while for the larger rings the interactions are expected to be more important.

4.7 Nanoribbon Rings

Graphene nanoribbon rings are graphene nanostructures which are formed by joining the two ends of a nanoribbon to form a ring [36]. The number of twists applied to the ribbon before the ends are attached changes the topology of the nanoribbon ring. The simplest ring is a ring with no twist which we will call a cyclic ring, shown in Fig. 4.37a. A more interesting and topologically different case occurs however when the ring is obtained by applying one twist to the ribbon before attaching the two ends, as shown in Fig. 4.37b. The resulting structure is called a Möbius ring, i.e., a surface with only one side and one edge. The Möbius ring is an interesting quantum dot; it has one edge like the dots studied up to now but its surface has no direction unlike flat 2D quantum dots studied in the previous section. It is also an example of a topological insulator, where the insulating behavior is generated by the finite width of the nanoribbon and nontrivial topology is realized explicitly through the Möbius twist [36, 60–67].

As we will see, for graphene structures with zigzag edges where the edge states play an important role, Möbius rings yield unusual electronic properties. Graphene nanoribbon rings along the armchair direction and/or with two or more twists can also be built, but in the following we will only consider zigzag-edged rings with one or no twists.

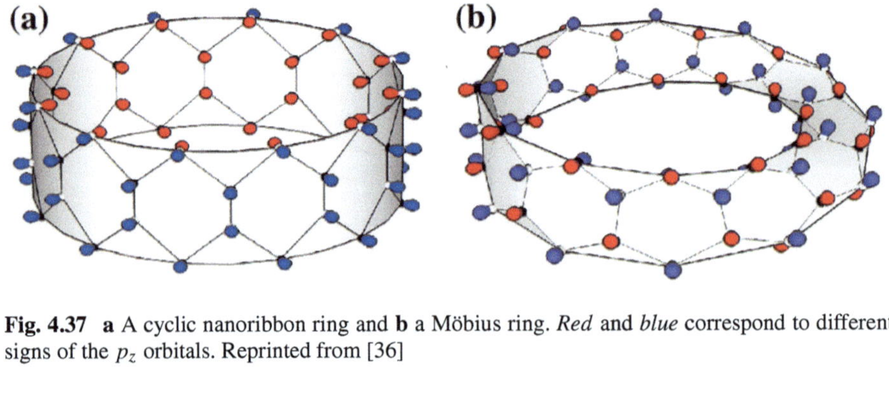

Fig. 4.37 **a** A cyclic nanoribbon ring and **b** a Möbius ring. *Red* and *blue* correspond to different signs of the p_z orbitals. Reprinted from [36]

4.7.1 Möbius and Cyclic Nanoribbon Rings

Before we describe a topologically nontrivial Möbius ring, let us start with the simpler cyclic ring. Figure 4.37 shows a cyclic nanoribbon ring with two zigzag chains. For a cyclic nanoribbon ring with N chains, the results presented in Sect. 4.2.3 apply directly. Hence, neglecting the spin-orbit coupling and ignoring the spin degree of freedom, we have:

$$H_k = U + \left(Te^{i4\pi k/M} + \text{h.c.}\right) \tag{4.53}$$

with

$$U = \begin{pmatrix} 0 & t & 0 & 0 & 0 & 0 & \ldots \\ t & 0 & t & 0 & 0 & 0 & \\ 0 & t & 0 & t & 0 & 0 & \\ 0 & 0 & t & 0 & t & 0 & \\ 0 & 0 & 0 & t & 0 & t & \\ 0 & 0 & 0 & 0 & t & 0 & \\ \ldots & & & & & & \ldots \end{pmatrix}; \quad T = \begin{pmatrix} 0 & 0 & 0 & 0 & 0 & 0 & \ldots \\ t & 0 & 0 & 0 & 0 & 0 & \\ 0 & 0 & 0 & t & 0 & 0 & \\ 0 & 0 & 0 & 0 & 0 & 0 & \\ 0 & 0 & 0 & 0 & 0 & 0 & \\ 0 & 0 & 0 & 0 & t & 0 & \\ \ldots & & & & & & \ldots \end{pmatrix}, \tag{4.54}$$

where N is the number of zigzag chains in the ribbon, M is the number of atoms in a single chain (as opposed to the number of unit cells defined in Sect. 4.2.3), $k = 0, \ldots, M/2 - 1$, and the size of the U and T matrices is $2N$. Alternatively, instead of going into the k-space representation, since we have a finite-size structure, one can build the tight-binding Hamiltonian of the size $NM \times NM$ and diagonalize it directly. This direct approach can be more practical for studying Coulomb interaction effects, but does not allow us to understand the effect of topology on the energy spectrum.

For the Möbius ring, the tight-binding Hamiltonian should be constructed in such a way that opposite corners of the two edges are connected. For p_z orbitals, additional care must be taken since during the twist the sign of the orbitals flips at the connection, changing the sign of the hopping parameter t. Further discussion on the Möbius boundary condition applied to graphene rings can be found in [36].

Figure 4.38 shows the energy spectrum for a $N = 2$, $M = 26$ ring, in Möbius and cyclic configurations. Since the total number of electrons is equal to the number of sites, we have $N_e = 52$ electrons filling the first 26 valence levels assumed for now to be doubly occupied. Note that cyclic configuration has the electron-hole symmetry, which is absent in the Möbius configuration. This broken electron-hole symmetry in the Möbius configuration has a subtle but important implication for wide ribbons. Figure 4.39 is a similar graph to Fig. 4.38 but for a wider ribbon with $N = 14$. For a ribbon with length M, as the width N increases, edge states become more distinguishable in the energy spectrum; their energies become increasingly degenerate at the Fermi level, with a substantial energy gap separating them from the remaining valence and conduction levels. For a ribbon ring in the cyclic configuration with a degenerate band of N_d edge states, charge neutrality requires the degenerate

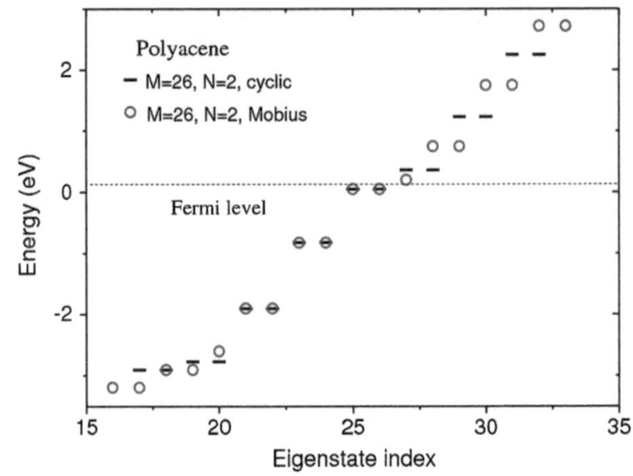

Fig. 4.38 Single particle energy spectrum of polyacene ring made of $N = 2$ zigzag chains, each containing $M = 26$ atoms (total of $N_a = 52$ atoms),in cyclic (*solid line*) and Möbius (*circles*) configurations. Cyclic and Möbius configurations share same valence band edge state energy levels. Reprinted from [36]

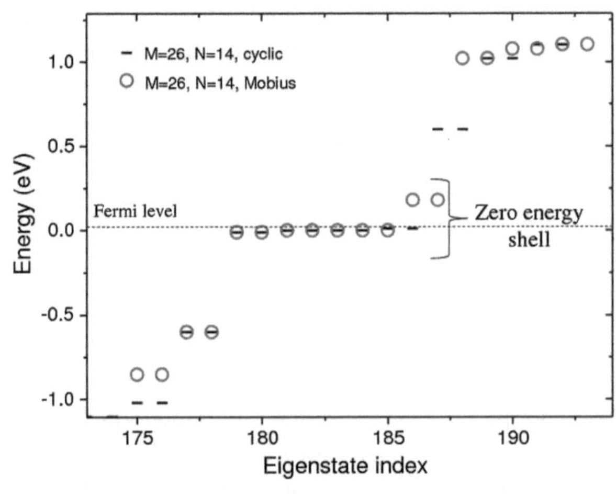

Fig. 4.39 Single particle energy spectrum of nanoribbon ring made of $N = 14$ zigzag chains, each containing $M = 26$ atoms (total of $N_a = 364$ atoms), in cyclic (*solid line*) and Möbius (*circles*) configurations. Möbius configuration (*circles*) has nine degenerate edge states occupied by eight electrons. Reprinted from [36]

band to be occupied with N_d electrons, leading to antiferromagnetic edge states [68, 69]. In the example given in Fig. 4.39, in the cyclic ring 8 electrons occupy 8 edge states giving a filling factor $\nu = 1$. The remaining electrons doubly occupy the valence states. However, for the Möbius configuration the situation is different. Due to the broken electron-hole symmetry we have 9 degenerate edge states occupied by

8 electrons, thus $\nu \neq 1$. As we will see in Chap. 6, the difference in the filling factor gives rise to different magnetic properties due to electron-electron interactions.

References

1. P.R. Wallace, Phys. Rev. **71**, 622 (1947)
2. L. Jacak, P. Hawrylak, A. Wojs, *Quantum Dots* (Springer, Berlin, 1998)
3. P. Hawrylak, M. Korkusinski, Electronic and optical properties of self-assembled quantum dots, in *Single Quantum Dots: Fundamentals, Applications, and New Concepts*, vol. 90, Topics in Applied Physics, ed. by P. Michler (Springer, Berlin, 2003), pp. 25–92
4. W. Sheng, M. Korkusinski, A.D. Guclu, M. Zielinski, P. Potasz, E.S. Kadantsev, O. Voznyy, P. Hawrylak, Frontiers Phys. **7**, 328 (2012)
5. S. Raymond, S. Studenikin, A. Sachrajda, Z. Wasilewski, S.J. Cheng, W. Sheng, P. Hawrylak, A. Babinski, M. Potemski, G. Ortner, M. Bayer, Phys. Rev. Lett. **92**, 187402 (2004)
6. A.H.C. Neto, F. Guinea, N.M.R. Peres, K.S. Novoselov, A.K. Geim, Rev. Mod. Phys. **81**, 109 (2009)
7. M.V. Berry, R.J. Mondragon, Proc. Royal Soc. Lond. A **412**, 53–74 (1987)
8. S. Schnez, K. Ensslin, M. Sigrist, T. Ihn, Phys. Rev. B **78**, 195427 (2008)
9. B. Wunsch, T. Stauber, F. Guinea, Phys. Rev. B **77**, 035316 (2008)
10. W. Husler, R. Egger, Phys. Rev. B **80**, 161402(R) (2009)
11. H.P. Heiskanen, M. Manninen, J. Akola, New J. Phys. **10**, 103015 (2008)
12. J. Akola, H.P. Heiskanen, M. Manninen, Phys. Rev. B **77**, 193410 (2008)
13. M. Zarenia, A. Chaves, G.A. Farias, F.M. Peeters, Phys. Rev. B **84**, 245403 (2011)
14. A.V. Rozhkov, F. Nori, Phys. Rev. B **81**, 155401 (2010)
15. M. Wimmer, A.R. Akhmerov, F. Guinea, Phys. Rev. B **82**, 045409 (2010)
16. F.E. Borghis, C.H. Papas, in *Encyclopedia of Physics*, ed. by S. Flucke (Springer, Berlin, 1957)
17. H.R. Krishnamurthy, H.S. Mani, H.C. Verma, J. Phys. A: Math. Gen. **15**, 2131 (1982)
18. J.W. McClure, Phys. Rev. **108**, 612 (1957)
19. Z.Z. Zhang, K. Chang, F.M. Peeters, Phys. Rev. B **77**, 235411 (2008)
20. D.A. Bahamon, A.L.C. Pereira, P.A. Schulz, Phys. Rev. B **79**, 125414 (2009)
21. N.M.R. Peres, J.N.B. Rodrigues, T. Stauber, J.M.B. Lopes dos Santos, J. Phys.: Condens. Matter **21**, 344202 (2009)
22. P. Recher, B. Trauzettel, Nanotechnology **21**, 302001 (2010)
23. F. Libisch, S. Rotter, J. Gttinger, C. Stampfer, J. Burgdrfer, Phys. Rev. B **81**, 245411 (2010)
24. M. Grujic, M. Zarenia, A. Chaves, M. Tadic, G.A. Farias, F.M. Peeters, Phys. Rev. B **84**, 205441 (2011)
25. A.D. Güçlü, P. Potasz, P. Hawrylak, Phys. Rev. B **88**, 155429 (2013)
26. S. Schnez, F. Molitor, C. Stampfer, J. Gttinger, I. Shorubalko, T. Ihn, K. Ensslin, Appl. Phys. Lett. **94**, 012107 (2009)
27. D.R. Hofstadter, Phys. Rev. B **14**, 2239 (1976)
28. Y. Hasegawa, P. Lederer, T.M. Rice, P.B. Wiegmannt, Phys. Rev. Lett. **63**, 907 (1989)
29. G. Dresselhaus, M.S. Dresselhaus, Phys. Rev. **140**, A401 (1965)
30. S. Konschuh, M. Gmitra, J. Fabian, Phys. Rev. B **82**, 245412 (2010)
31. C.L. Kane, E.J. Mele, Phys. Rev. Lett. **95**, 226801 (2005)
32. R. Saito, G. Dresselhaus, M.S. Dresselhaus, *Physical Properties of Carbon Nanotubes* (Imperial College Press, London, 1998)
33. Manuel J. Schmidt, Daniel Loss, Phys. Rev. B **81**, 165439 (2010)
34. D.J. Chadi, Phys. Rev. B **16**, 790 (1977)
35. M. Zielinski, M. Korkusiski, P. Hawrylak, Phys. Rev. B **81**, 085301 (2010)
36. A.D. Guclu, M. Grabowski, P. Hawrylak, Phys. Rev. B **87**, 035435 (2013)
37. K. Nakada, M. Fujita, G. Dresselhaus, M.S. Dresselhaus, Phys. Rev. B **54**, 17954 (1996)

38. M.Z. Hasan, C.L. Kane, Rev. Mod. Phys. **82**, 3045 (2010)
39. Y.S. Dedkov, M. Fonin, U. Rdiger, C. Laubschat, Phys. Rev. Lett. **100**, 107602 (2008)
40. M. Gmitra, S. Konschuh, C. Ertler, C. Ambrosch-Draxl, J. Fabian, Phys. Rev. B **80**, 235431 (2009)
41. D. Huertas-Hernando, F. Guinea, A. Brataas, Phys. Rev. B **74**, 155426 (2006)
42. F. Kuemmeth, S. Ilani, D.C. Ralph, P.L. McEuen, Nature (London) **452**, 448 (2008)
43. P. Potasz, A.D. Güçlü, P. Hawrylak, Phys. Rev. B **81**, 033403 (2010)
44. E.W. Weisstein, Pascals Triangle. From MathWorldAWolfram Web Resource. http://mathworld.wolfram.com/PascalsTriangle.html
45. M.S. Dresselhaus, G. Dresselhaus, A. Jorio, *Group Theory: Application to the Physics of Condensed Matter* (Springer, Berlin, 2008)
46. Y.W. Son, M.L. Cohen, S.G. Louie, Phys. Rev. Lett. **97**, 216803 (2006)
47. O.V. Yazyev, R.B. Capaz, S.G. Louie, Phys. Rev. B **84**, 115406 (2011)
48. C. Tao, L. Jiao, O.V. Yazyev, Y.-C. Chen, J. Feng, X. Zhang, R.B. Capaz, J.M. Tour, A. Zettl, S.G. Louie, H. Dai, M.F. Crommie, Nat. Phys. **7**, 616 (2011)
49. A.D. Güçlü, P. Potasz, P. Hawrylak, Phys. Rev. B **84**, 035425 (2011)
50. J. Fernandez-Rossier, J.J. Palacios, Phys. Rev. Lett. **99**, 177204 (2007)
51. W.L. Wang, S. Meng, E. Kaxiras, Nano Lett. **8**, 241 (2008)
52. M. Ezawa, Phys. Rev. B **77**, 155411 (2008)
53. A.D. Güçlü, P. Potasz, O. Voznyy, M. Korkusinski, P. Hawrylak, Phys. Rev. Lett. **103**, 246805 (2009)
54. A.D. Güçlü, P. Potasz, P. Hawrylak, Phys. Rev. B **82**, 155445 (2010)
55. M. Ezawa, Phys. Rev. B **81**, 201402 (2010)
56. O. Voznyy, A.D. Güçlü, P. Potasz, P. Hawrylak, Phys. Rev. B **83**, 165417 (2011)
57. P. Potasz, A.D. Güçlü, O. Voznyy, J.A. Folk, P. Hawrylak, Phys. Rev. B **83**, 174441 (2011)
58. P. Potasz, A.D. Güçlü, P. Hawrylak, Phys. Rev. B **82**, 075425 (2010)
59. E.H. Lieb, Phys. Rev. Lett. **62**, 1201 (1989)
60. R. Herges, Chem. Rev. **106**, 4820 (2006)
61. D. Ajami, O. Oeckler, A. Simon, R. Herges, Nature **426**, 819 (2003)
62. K. Wakabayashi, K. Harigaya, J. Phys. Soc. Jpn. **72**, 998 (2003)
63. K. Harigaya, A. Yamashiro, Y. Shimoi, K. Wakabayashi, Synth. Met. **152**, 261 (2005)
64. D. Jiang, S. Dai, J. Phys. Chem. C **112**, 5348 (2008)
65. E.W.S. Caetano, V.N. Freire, S.G. dos Santos, D.S. Galvao, F. Sato, J. Chem. Phys. **128**, 164719 (2008)
66. Z.L. Guo, Z.R. Gong, H. Dong, C.P. Sun, Phys. Rev. B **80**, 195310 (2009)
67. X. Wang, X. Zheng, M. Ni, L. Zou, Z. Zeng, Appl. Phys. Lett. **97**, 123103 (2010)
68. A. Yamashiro, Y. Shimoi, K. Harigaya, K. Wakabayashi, Phys. Rev. B **68**, 193410 (2003)
69. K. Harigaya, A. Yamashiro, Y. Shimoi, K. Wakabayashi, Synth. Met. **152**, 317 (2005)

Chapter 5
Electron–Electron Interactions in Graphene Quantum Dots

Abstract This chapter introduces the problem of electron–electron interactions, briefly describes several methods and their application to graphene quantum dots. The Hubbard model, the mean-field Hartree-Fock method, the Density Functional Theory and the configuration interaction (CI) method are introduced and applied to graphene quantum dots.

5.1 Introduction

The problem of electron–electron interactions in condensed matter remains a challenge described in many excellent references [1–3]. Some progress has been made in artificially structured two-, one- and zero-dimensional materials where the effects of electron–electron interactions could be studied in a controlled way. This includes 2D and layered electron gases [4, 5], semiconductor quantum dots with controlled electron numbers [6, 7], and graphene. The electron–electron interactions in graphene and multi-layer graphene have been extensively studied, starting with intercalated graphite [8–10], with results of recent work reviewed in a number of excellent monographs [11–18].

There are several approaches to the problem of interacting electrons in graphene. The most common starts with the effective mass approximation for the tight-binding band structure in the form of two-dimensional Dirac electrons in the two nonequivalent valleys. The electron–electron interaction is treated as a three-dimensional Coulomb interaction of Dirac Fermions [11, 13, 14]. In this approach the Dirac form of single particle spectrum, with relativistic dispersion of electrons and holes plays an important role, with strong analogies to Quantum Electrodynamics (QED), and, e.g., logarithmically divergent exchange self-energy, instability due to spontaneous electron-hole pair formation (excitonic instability) and atomic collapse resonances [11, 13, 14, 19].

One can start to appreciate the different role of electron–electron interactions for Schrödinger and Dirac Fermions by comparing the two effective mass Hamiltonians.

For 2D Schrödinger electrons with a parabolic dispersion and the effective mass m^*, the Hamiltonian reads ($\hbar = 1$):

A.D. Güçlü et al., *Graphene Quantum Dots*,
NanoScience and Technology, DOI 10.1007/978-3-662-44611-9_5

$$H_{SE} = \sum_i \left(\frac{(-i\nabla_i)^2}{2m^*}\right) + \frac{1}{2}\sum_{i,j}\frac{e^2}{\kappa|\mathbf{r}_i - \mathbf{r}_j|}, \tag{5.1}$$

while for 2D Dirac electrons, 2.24, interacting with the same 3D Coulomb interactions screened by the dielectric constant κ, the Hamiltonian reads:

$$H_{DE} = \sum_i ((3|t|a/2)\sigma \cdot (-i\nabla_i)) + \frac{1}{2}\sum_{i,j}\frac{e^2}{\kappa|\mathbf{r}_i - \mathbf{r}_j|}. \tag{5.2}$$

Scaling first the energy in effective Rydbergs $Ry = \frac{1}{2m^* a_0^2} = \frac{e^2}{2\kappa a_0}$ and distance in Bohr radius a_0 and then, scaling the length r in the average distance between electrons, r_s, given by the two-dimensional electron density $n = \frac{1}{\pi r_s^2}$, allows us to rewrite the two Hamiltonians as:

$$H_{SE} = \frac{Ry}{r_s^2}\left[\sum_i (-i\nabla_i)^2 + (r_s)\frac{1}{2}\sum_{i,j}\frac{2}{\kappa|\mathbf{r}_i - \mathbf{r}_j|}\right], \tag{5.3}$$

and

$$H_{DE} = \frac{Ry3|t|a}{2r_s}\left[\sum_i -i(\sigma \cdot \nabla_i) + \left(\frac{1}{3|t|a/2}\right)\frac{1}{2}\sum_{i,j}\frac{2}{\kappa|\mathbf{r}_i - \mathbf{r}_j|}\right]. \tag{5.4}$$

We see that for Schrödinger electrons the electron–electron interaction term is proportional to r_s and hence its contribution can be made arbitrarily small for small r_s, i.e., high electron density. For high, but finite electron density the perturbation theory can be applied to calculate the effects of electron–electron interactions. By contrast, for Dirac electrons the contribution of electron–electron interactions to the total energy is independent of r_s and carrier density. The electron–electron contribution depends inversely on the tunneling matrix element and lattice constant, measured in effective, screened Rydberg and the screened lattice constant, or Fermi velocity. The more itinerant, mobile, the electrons are, the weaker the effect of electron–electron interactions.

The density dependence of electron–electron interactions is brought back to graphene by the application of a large perpendicular magnetic field. The filling-factor-dependent effects of the electron–electron interaction in the integer and fractional quantum Hall effect are treated in the effective mass approach [20–22]. In the fractional quantum Hall regime, the configuration interaction (CI) methods for electrons on a Haldane sphere or torus are used to understand electron–electron effects in the degenerate shells of Landau levels in graphene [20–22].

In the second approach graphene is treated as any other solid, a collection of nuclei and electrons. The ground-state energy and density is evaluated using ab initio density functional methods. The Kohn-Sham quasiparticles are used as a starting point for the many-body perturbation theory (GW) [11, 12, 15, 23]. Optical properties are calculated solving the Bethe-Salpeter equation for the excited quasi-electron and the quasi-hole which is left behind. Ab initio approaches are important for determining the role of adsorbates, defects and edges in graphene [19, 24–27].

The third approach treats graphene as a lattice of sites hosting p_z orbitals and adds electron–electron interactions following the extended Hubbard model. Here, either mean-field and/or Hartree-Fock approaches are used or Quantum Monte Carlo is applied to determine the ground-state properties. For example, Sorella and co-workers applied QMC technique to establish the range of Hubbard parameters corresponding to the semi-metallic ground state well described by the tight-binding model where tight-binding parameters include electron–electron interactions at the mean-field level. For stronger Coulomb interactions a transition to Mott insulator is predicted [16, 28–30].

These different methodologies have also been applied to electron–electron interactions in graphene quantum dots and are described in some detail in the following chapters.

5.2 Many-Body Hamiltonian

The starting point for the understanding of graphene quantum dots is the Hamiltonian of interacting electrons, each moving in a field of attractive potential V_{ion} of nuclei:

$$H_{MB} = \sum_i \left(-\nabla_i^2 + \sum_j V_{ion}^j(\mathbf{r}_i - \mathbf{R}_j) \right) + \frac{1}{2} \sum_{i,j} \frac{2}{|\mathbf{r}_i - \mathbf{r}_j|}. \tag{5.5}$$

Here $V_{ion}^j(\mathbf{r} - \mathbf{R}) = -\frac{2}{|\mathbf{r}-\mathbf{R}|}$ is the potential produced by the positively charged nucleus at $\mathbf{R}$ and the second term describes Coulomb interactions of all electrons, 6 per carbon atom, written in Rydbergs and Bohr radii.

However, we are mainly interested in either valence electrons, or p_z electrons only. For single-particle Hamiltonian describing only p_z electrons the nuclear potential, V_{ion}, is screened by core and sigma electrons and is replaced by a pseudo-potential V_{eff}. One often also includes screening by a surrounding medium by introducing screened Coulomb electron–electron interactions, with the effective Hamiltonian for p_z electrons written as:

$$H_{MB} = \sum_i \left(-\nabla_i^2 + \sum_j V_{eff}^j(\mathbf{r}_i - \mathbf{R}_j) \right) + \frac{1}{2} \sum_{i,j} \frac{2}{\kappa|\mathbf{r}_i - \mathbf{r}_j|}, \tag{5.6}$$

with κ being the dielectric constant. We note that one implicitly assumes that the surrounding medium cannot screen the ionic potential but does screen the long range electron–electron interaction [13, 30].

The many-body Hamiltonian can be rewritten in second quantization. We first establish the orthogonal single-particle basis by, e.g., choosing eigenstates $\phi_j(\mathbf{r})$ of the atomic Hamiltonian in (5.6), and orthogonalizing them with respect to different atoms.

We next expand the field operators $\Phi(\mathbf{r}) = \sum_j c_j \phi_j(\mathbf{r})$ in terms of basis states and creation/annihilation operators, $c_{i\sigma}$ ($c^{\dagger}_{i\sigma}$) which annihilate (create) an electron on ith p_z orbital with spin σ. These operators satisfy anticommutation rules, $\{c_{i\sigma}, c_{j\sigma}\} = \{c^{\dagger}_{i\sigma}, c^{\dagger}_{j\sigma}\} = 0$ and $\{c_{i\sigma'}, c^{\dagger}_{j\sigma}\} = \delta_{ij}\delta_{\sigma\sigma'}$, which guarantee the antisymmetry of many-body states. The Hamiltonian given by (5.6) may be written in the second quantization form as [1–3, 31]

$$H = \sum_{i,\sigma} \varepsilon_{i\sigma} c^{\dagger}_{i\sigma} c_{i\sigma} + \sum_{i,l,\sigma} \tau_{il\sigma} c^{\dagger}_{i\sigma} c_{l\sigma} + \frac{1}{2} \sum_{\substack{i,j,k,l,\\ \sigma\sigma'}} \langle ij|V|kl\rangle c^{\dagger}_{i\sigma} c^{\dagger}_{j\sigma'} c_{k\sigma'} c_{l\sigma}, \tag{5.7}$$

with tunneling matrix elements $\tau_{il\sigma}$ defined by (2.14) and Coulomb matrix elements $\langle ij|V|kl\rangle$ described in detail in the next subsection. The first term in (5.7) corresponds to the energy of p_z orbitals, $\varepsilon_{i\sigma} = \varepsilon$.

5.3 Two-body Scattering—Coulomb Matrix Elements

The two-body Coulomb term from (5.7) is written as

$$\frac{1}{2} \sum_{\substack{i,j,k,l,\\ \sigma\sigma'}} \langle ij|V|kl\rangle c^{\dagger}_{i\sigma} c^{\dagger}_{j\sigma'} c_{k\sigma'} c_{l\sigma}, \tag{5.8}$$

with Coulomb matrix elements, in Rydbergs, defined as

$$\langle ij \mid V \mid kl\rangle = \int\int d\mathbf{r}_1 d\mathbf{r}_2 \phi^*_i(\mathbf{r}_1)\, \phi^*_j(\mathbf{r}_2) \frac{2}{\kappa \mid \mathbf{r}_2 - \mathbf{r}_1 \mid} \phi_k(\mathbf{r}_2)\, \phi_l(\mathbf{r}_1), \tag{5.9}$$

where $\mathbf{r}_1$ and $\mathbf{r}_2$ are coordinates of the first and the second electron, respectively. The Coulomb matrix element describes scattering of two electrons occupying orbitals on sites with indices k and l to two orbitals on sites with indices j and i. Note that for $l = i$ and $k = j$, $\langle ij|V|ji\rangle$ corresponds to the Coulomb interaction between two electronic densities localized on sites i and j. On the other hand, for $k = i$ and $l = j$, $\langle ij|V|ij\rangle$ corresponds to the exchange term which appears only when electrons on i and j orbitals have the same spin, $\sigma = \sigma'$ in (5.9).

Table 5.1 Selected Coulomb matrix elements between electrons on sites in graphene honeycomb lattice for $\kappa = 1$

$\langle ij \mid V \mid kl \rangle$	E (eV)
$\langle 11 \mid V \mid 11 \rangle$	16.522
$\langle 12 \mid V \mid 21 \rangle$	8.640
$\langle 13 \mid V \mid 31 \rangle$	5.333
$\langle 11 \mid V \mid 12 \rangle$	3.157
$\langle 12 \mid V \mid 31 \rangle$	1.735
$\langle 12 \mid V \mid 12 \rangle$	0.873
$\langle 11 \mid V \mid 22 \rangle$	0.873
$\langle 22 \mid V \mid 13 \rangle$	0.606

Numbers 1, 2 and 3 indicate electron on-site, on nearest-neighbor site and on next-nearest-neighbor site of hexagonal lattice, respectively

The p_z orbitals of carbon atoms given in (5.9) can be approximated by Slater orbitals, given by

$$\phi_i(\mathbf{r}_1) = \left(\frac{\xi^5}{32\pi}\right)^{\frac{1}{2}} z \exp\left(\frac{-\xi r_1}{2}\right), \tag{5.10}$$

with $\xi = 3.14$ [32]. Coulomb matrix elements given by (5.9) can be numerically calculated for orbitals localized on lattice sites of the honeycomb graphene lattice [33, 34]. In our numerical calculations, on-site, scattering, and exchange terms up to the next-nearest neighbors, as well as all long-range direct terms were obtained. In Table 5.1 we show selected Coulomb matrix elements for dielectric constant $\kappa = 1$. Numbers 1, 2 and 3 indicate electron on-site, on nearest-neighbor site and on next-nearest-neighbor site of the honeycomb graphene lattice, respectively.

5.4 Mean-Field Hartree-Fock Approximation

The many-body Hamiltonian cannot be solved but for a few electrons on a few orbitals. Hence, the electron–electron scattering two-body term is replaced by a one electron moving in a mean field of other electrons, to be determined self-consistently. One starts by replacing one of the electrons in the two-body Coulomb interaction term from (5.7) by its expectation value: $c^\dagger_{i\sigma} c^\dagger_{j\sigma'} c_{k\sigma'} c_{l\sigma} \simeq c^\dagger_{i\sigma} \langle c^\dagger_{j\sigma'} c_{k\sigma'} \rangle c_{l\sigma} - c^\dagger_{i\sigma} \langle c^\dagger_{j\sigma'} c_{l\sigma} \rangle c_{k\sigma'}$ where $\langle c^\dagger_{j\sigma'} c_{k\sigma'} \rangle$ is an element of the density matrix for a pair of states j, k, to be determined. There are four ways to pair creation and annihilation operators, allowing us to write the Coulomb operator from (5.7) in the mean-field approximation as

$$
\begin{aligned}
V_{MF} = & \frac{1}{2} \sum_{\substack{i,j,k,l,\\ \sigma\sigma'}} \langle ij|V|kl\rangle \left(\langle c^{\dagger}_{j\sigma'} c_{k\sigma'}\rangle c^{\dagger}_{i\sigma} c_{l\sigma} + \langle c^{\dagger}_{i\sigma} c_{l\sigma}\rangle c^{\dagger}_{j\sigma'} c_{k\sigma'} \right) \\
& - \frac{1}{2} \sum_{\substack{i,j,k,l,\\ \sigma\sigma'}} \langle ij|V|kl\rangle \left(\langle c^{\dagger}_{j\sigma'} c_{l\sigma}\rangle c^{\dagger}_{i\sigma} c_{k\sigma'} \delta_{\sigma\sigma'} + \langle c^{\dagger}_{i\sigma} c_{k\sigma'}\rangle c^{\dagger}_{j\sigma'} c_{l\sigma} \delta_{\sigma\sigma'} \right), \quad (5.11)
\end{aligned}
$$

where the first part corresponds to direct terms and the second part to exchange terms.

Combining and rearranging the two direct and two exchange terms cancels the factor of $1/2$ and allows us to write the mean-field Coulomb terms as:

$$
V_{MF} = \sum_{\substack{i,j,k,l,\\ \sigma\sigma'}} \left(\langle ij|V|lk\rangle - \langle ij|V|kl\rangle \delta_{\sigma\sigma'} \right) \langle c^{\dagger}_{j\sigma'} c_{k\sigma'}\rangle c^{\dagger}_{i\sigma} c_{l\sigma} .
$$

Finally, the Hamiltonian given by (5.7) can be written in the mean-field HF approximation as

$$
\begin{aligned}
H_{MF} = & \sum_{i,\sigma} \varepsilon_{i\sigma} c^{\dagger}_{i\sigma} c_{i\sigma} + \sum_{i,l,\sigma} \tau_{il\sigma} c^{\dagger}_{i\sigma} c_{l\sigma} \\
& + \sum_{\substack{i,l,\\ \sigma}} [\sum_{\substack{j,k,\\ \sigma'}} \left(\langle ij|V|kl\rangle - \langle ij|V|lk\rangle \delta_{\sigma\sigma'} \right) \langle c^{\dagger}_{j\sigma'} c_{k\sigma'}\rangle] c^{\dagger}_{i\sigma} c_{l\sigma} . \quad (5.12)
\end{aligned}
$$

The density matrix elements $\langle c^{\dagger}_{j\sigma'} c_{k\sigma'}\rangle$ need to be determined self-consistently by iterating (5.12).

5.4.1 Hartree-Fock State in Graphene Quantum Dots

In the previous section, the general form of the many-body Hamiltonian in the mean-field HF approximation, (5.12), was written. Before proceeding to graphene quantum dots we examine the HF state in bulk graphene. The HF Hamiltonian, (5.12), for a graphene layer can be written as

$$
\begin{aligned}
H^{o}_{MF} = & \sum_{i}^{\infty} \sum_{\sigma} \varepsilon_{i\sigma} c^{\dagger}_{i\sigma} c_{i\sigma} + \sum_{i,l}^{\infty} \sum_{\sigma} \tau_{il\sigma} c^{\dagger}_{i\sigma} c_{l\sigma} \\
& + \sum_{i,j,k,l}^{\infty} \sum_{\sigma\sigma'} \left(\langle ij|V|kl\rangle - \langle ij|V|lk\rangle \delta_{\sigma\sigma'} \right) \rho^{o}_{jk\sigma'} c^{\dagger}_{i\sigma} c_{l\sigma} \quad (5.13) \\
= & \sum_{i,l}^{\infty} \sum_{\sigma} t_{il\sigma} c^{\dagger}_{i\sigma} c_{l\sigma} , \quad (5.14)
\end{aligned}
$$

with $\rho^{o}_{jk\sigma'} = \langle ... \rangle_{GS}$ being the density matrix elements. This is effectively a one-body TB Hamiltonian given by (4.1) with the experimentally measured hopping integral t_{il} for graphene [35]. We proceed to evaluate density matrix elements with respect to the ground state (GS)- the fully occupied valence band of the TB Hamiltonian. The density matrix can be written as

$$\rho^{o}_{jk\sigma'} = \sum_{\mathbf{k}} b^{*}_{\mathbf{R}_j}(\mathbf{k}) b_{\mathbf{R}_k}(\mathbf{k}), \tag{5.15}$$

where j and k are graphene lattice sites and the summation is over the full valence band. $b_{\mathbf{R}}$'s are the coefficients of the p_z orbitals which according to (2.17) can be written as

$$b_{\mathbf{R}_j} = \frac{1}{\sqrt{2N_c}} e^{i\mathbf{k}\cdot\mathbf{R}_j}, \tag{5.16}$$

for A-type atoms and

$$b_{\mathbf{R}_j} = \frac{1}{\sqrt{2N_c}} e^{i\mathbf{k}\cdot\mathbf{R}_j} e^{-i\theta_{\mathbf{k}}}, \tag{5.17}$$

for B-type atoms. Due to the translational invariance of graphene, the density matrix depends only on relative positions $|\mathbf{R_j} - \mathbf{R_k}|$. On-site density matrix elements for an arbitrary lattice site j are site and sublattice index independent,

$$\rho^{o}_{jj\sigma'} = \frac{1}{2N_c} \sum_{\mathbf{k}} e^{-i\mathbf{k}\cdot\mathbf{R}} e^{i\mathbf{k}\cdot\mathbf{R}} = \frac{1}{2N_c} \sum_{\mathbf{k}} 1 = \frac{1}{2}, \tag{5.18}$$

where we took into account the fact that the number of occupied states is equal to the number of unit cells in the system. The number of all energy levels is $2N_c$ with two atoms in the unit cell, and only half of them are occupied, such that the summation in (5.18) gives N_c. The nearest-neighbor density matrix elements for atoms from the same unit cell correspond to $\mathbf{R_k} = \mathbf{R_j}$ and are evaluated using

$$\rho^{o}_{jk\sigma'} = \frac{1}{2N_c} \sum_{\mathbf{k}} e^{-i\mathbf{k}\cdot\mathbf{R}_j} e^{i\mathbf{k}\cdot\mathbf{R}_k} e^{-i\theta_{\mathbf{k}}} = \frac{1}{2N_c} \sum_{\mathbf{k}} e^{-i\theta_{\mathbf{k}}} \simeq 0.262,$$

where the summation over occupied valence states is carried out numerically. We note that one obtains the same value for two other nearest-neighbors. Same results can also be obtained by diagonalizing the tight-binding Hamiltonian for a sufficiently large graphene quantum dot, and by computing the density matrix elements for two nearest neighbors in the vicinity of the center of the structure. We have also calculated next-nearest neighbors density matrix elements, getting a negligibly small value.

For graphene quantum dots one can start directly with the mean-field HF Hamiltonian. An alternative is to use the Hamiltonian in a mean-field approximation starting from bulk HF single particle energy levels obtained within the HF-TB model. In order to do this, we combine (5.12) and (5.14) obtaining

$$
\begin{aligned}
H_{MF}^{GQD} &= H_{MF}^{GQD} - H_{MF}^{o} + H_{MF}^{o} \\
&= \sum_{i}^{N}\sum_{\sigma} \varepsilon_{i\sigma} c_{i\sigma}^{\dagger} c_{i\sigma} + \sum_{i,l}^{N}\sum_{\sigma} t_{il\sigma} c_{i\sigma}^{\dagger} c_{l\sigma} \\
&+ \sum_{i,l}^{N}\sum_{\sigma\sigma'}\sum_{j,k}^{N} \rho_{jk\sigma'} (\langle ij|V|kl\rangle - \langle ij|V|lk\rangle \delta_{\sigma,\sigma'}) c_{i\sigma}^{\dagger} c_{l\sigma}, \\
&- \sum_{i,l}^{N}\sum_{\sigma\sigma'}\sum_{j,k}^{\infty} \rho_{jk\sigma'}^{o} (\langle ij|V|kl\rangle - \langle ij|V|lk\rangle \delta_{\sigma,\sigma'}) c_{i\sigma}^{\dagger} c_{l\sigma}.
\end{aligned}
\tag{5.19}
$$

Note that the summation over j and k in the last term extends to infinity. For simplicity, in the following we will limit the summation to N, i.e., to quantum dot sites. This means that we are neglecting the three- and four-center scattering and exchange integrals nearby the edges of the quantum dot which are small. We finally obtain an approximate mean-field quantum dot Hamiltonian:

$$
\begin{aligned}
H_{MF}^{GQD} &= \sum_{i}^{N}\sum_{\sigma} \varepsilon_{i\sigma} c_{i\sigma}^{\dagger} c_{i\sigma} + \sum_{i,l}^{N}\sum_{\sigma} t_{il\sigma} c_{i\sigma}^{\dagger} c_{l\sigma} \\
&+ \sum_{i,l}^{N}\sum_{\sigma\sigma'}\sum_{j,k}^{N} (\rho_{jk\sigma'} - \rho_{jk\sigma'}^{o}) (\langle ij|V|kl\rangle - \langle ij|V|lk\rangle \delta_{\sigma,\sigma'}) c_{i\sigma}^{\dagger} c_{l\sigma}.
\end{aligned}
\tag{5.20}
$$

The density matrix elements $\rho_{jk\sigma'}$ are calculated with respect to the many-body ground state of graphene nanostructures. They can be written as

$$
\rho_{jk} = \sum_{\mathbf{s}} (A_j^s)^* A_k^s, \tag{5.21}
$$

where indices s run over all occupied states and A_j^s are expansion coefficients of eigenstates written in the basis of localized p_z orbitals

$$
c_s^{\dagger} = \sum_{i,\sigma} (A_i^s)^* a_i^{\dagger}.
$$

The Hamiltonian given by (5.20) has to be solved self-consistently to obtain Hartree-Fock quasi-particle orbitals.

In the following chapters we will use both approaches, the direct mean-field HF Hamiltonian of (5.12) and the approximate mean-field graphene quantum dot HF Hamiltonian, (5.20), with the density matrix measured from the bulk density matrix.

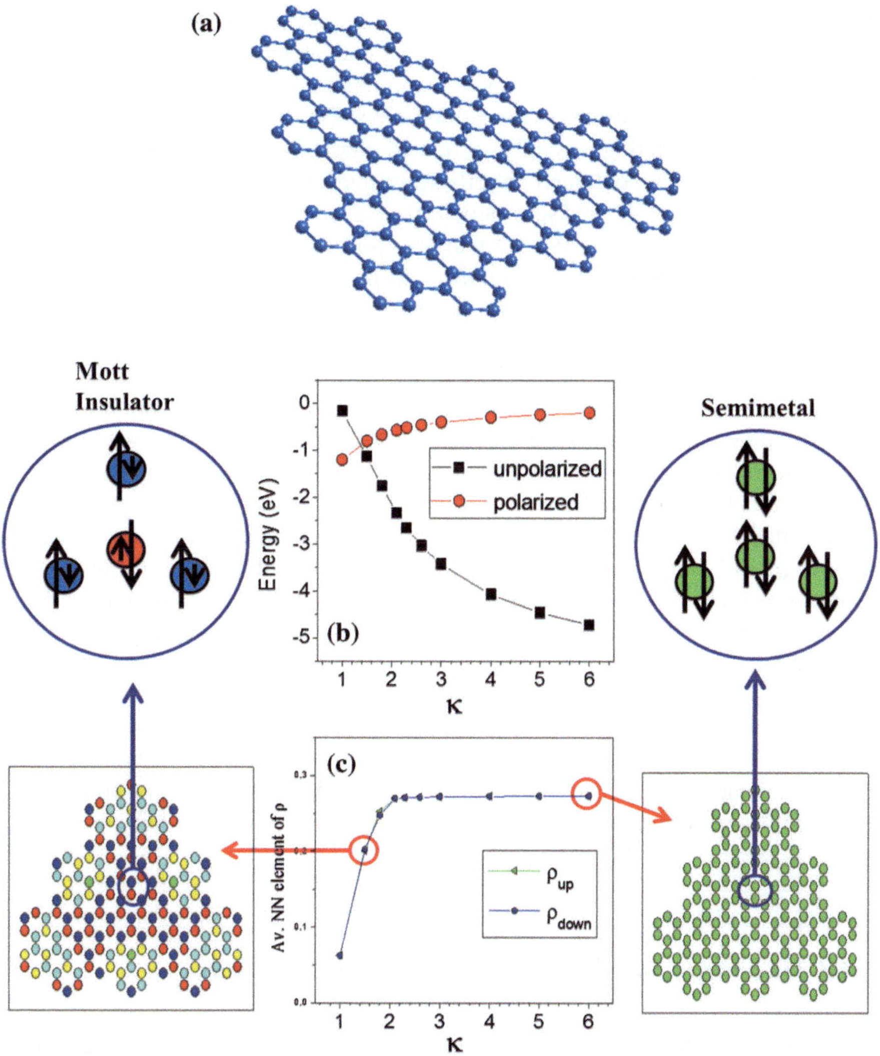

Fig. 5.1 **a** C168 Colloidal graphene quantum dot with 168 atoms. **b–c** Phase diagram of C168 at $t = -4.2\,\text{eV}$, $t_2 = -0.1\,\text{eV}$. **b** Ground state energy of the spin polarized and spin unpolarized C168 and **c** the nearest-neighbor density matrix element of the spin unpolarized C168 as a function of screening strength κ. Reprinted from [28]

5.4.2 Semimetal-Mott Insulator Transition in Graphene Quantum Dots

Following Ozfidan et al. in [28], we will now use the mean-field Hartree-Fock approximation described above to study the Mott transition in a quantum dot, shown in Fig. 5.1a [28], which has $N = 168$ carbon atoms and $N = 168$ p_z electrons.

Previous work on the ground-state properties of graphene [16, 28–30] suggests that the ground-state properties depend strongly on the values of the screening constant and the amplitude of the hopping term. In particular, for strong Coulomb interactions, or small values of κ, there exists a transition from a semi-metallic, weakly-interacting phase to a Mott-insulating, strongly correlated phase. Figure 5.1b shows the energy of the HF ground state obtained directly from the HF Hamiltonian, (5.12), for the spin-polarized, $S_z = N/2$, and spin-unpolarized, $S_z = 0$, states of C168 as a function of κ for $t = -4.2$ eV [28]. We see that the spin-unpolarized phase is the ground state for all κ down to $\kappa = 1.4$ while the spin-polarized state is the ground state at $\kappa < 1.4$, most likely an artifact of the Hartree-Fock approach. We now monitor the evolution of spin-unpolarized phase as a function of screening. Figure 5.1c shows the calculated average density matrix element $\rho_\sigma = \langle c^\dagger_{i\sigma} c_{j\sigma} \rangle$ for i, j nearest neighbors, averaged over all pairs for a spin-unpolarized ground state as a function of κ. The density matrix element shows the probability of having two electrons with the same spin on nearest-neighbor orbitals. For large κ we find $\rho_\sigma = 0.26$, i.e., the value for the HF state of bulk graphene, as discussed above. The local values of ρ_σ of course differ from the bulk value at the edges even in the range of high κ. As κ decreases and the strength of Coulomb interactions increases, we see the decrease in the probability of having two electrons with parallel spin on the two sublattices at around $\kappa < 1.8$. For $\kappa = 6$ and $\rho_\sigma = 0.26$ the right-hand side inset shows the spin density in the center of the quantum dot. We see that the carbon atom and its nearest neighbors are occupied by equally probable spin up and down electrons. Hence, the probability of finding a spin-up electron on the nearest neighbor atom is high for a spin-up electron in the center. For $\kappa < 1.5$ the ground state departs from the semiconducting state of the graphene quantum dot and becomes a Mott-insulator, with spin-up electrons on lattice A and spin-down electrons on lattice B as shown on the left hand side of Fig. 5.1.

5.4.3 Hubbard Model—Mean-Field Approximation

The Hubbard model [36–38] is frequently used to describe the effects of electron–electron interactions on the electronic properties of graphene quantum dots [39–41], in particular the spin polarization. In the Hubbard approximation all scattering matrix elements $\langle ij|V|kl\rangle$ are neglected except for the onsite terms $\langle ii|V|ii\rangle = U$, penalizing spin-up and down electrons occupying the same site i.

The Hubbard model in the mean-field approximation is described by the Hamiltonian:

$$H = \sum_{i,j} -t c^\dagger_{i\uparrow} c_{j\uparrow} + \sum_i U \langle c^\dagger_{i\downarrow} c_{i\downarrow} \rangle c^\dagger_{i\uparrow} c_{i\uparrow} + \sum_{i,j} -t c^\dagger_{i\downarrow} c_{j\downarrow} + \sum_i U \langle c^\dagger_{i\uparrow} c_{i\uparrow} \rangle c^\dagger_{i\downarrow} c_{i\downarrow}. \tag{5.22}$$

The Hamiltonian given by (5.22) consists of two blocks, for spin-up and spin-down states. Additionally, spin up Hamiltonian depends on spin down densities and vice versa. As a starting point, one can choose a simple TB Hamiltonian given by (4.1). Eigenvalues and eigenvectors are obtained after diagonalizing the TB Hamiltonian. The energy levels are filled by N_{dn} and N_{up} electrons, occupying N_{dn} and N_{up} lowest eigenstates, respectively. Next, spin-up and spin-down densities on each site can be calculated. According to (5.22), the calculated spin-down densities correspond to diagonal matrix elements of the spin-up Hamiltonian and calculated spin-up densities correspond to diagonal matrix elements of spin-down Hamiltonian. After diagonalizing separately spin-down and spin-up Hamiltonians, new energy levels for spin-down and spin-up electrons are obtained. These new states are again filled by N_{dn} and N_{up} electrons which occupy N_{dn} and N_{up} lowest energy levels, respectively. New spin densities can be calculated and used in new spin-up and spin-down Hamiltonians. The procedure is repeated until convergence with an appropriate accuracy is obtained.

5.5 Ab Inito Density Functional Approach

The many-body problem given by (5.6) cannot be solved but for very few electrons. However, the Density Functional Theory [42] (DFT) established that the ground state energy $E[n]$ is a functional of electron density $n(r)$ and not the many-body wavefunction which depends on coordinates of all electrons. The DFT replaces the many-body interacting problem by a system of noninteracting Kohn-Sham quasiparticles.

In DFT the total energy $E[n]$ can be written as

$$E[n] = T[n] + U[n] + \frac{1}{2}\int d\mathbf{r}' \frac{n(\mathbf{r})n(\mathbf{r}')}{|\mathbf{r}-\mathbf{r}'|} + E_{xc}[n], \tag{5.23}$$

where T is the kinetic energy and U is the confining potential energy. The last two terms on the right-hand side come from the electron–electron interactions and are written as the sum of the Hartree energy (third term) and the exchange-correlation energy E_{xc}. Note that E_{xc} has no explicit functional form and is defined through (5.23). Kohn and Sham [43] showed that $n(\mathbf{r})$ can be computed by solving the self-consistent set of equations (in atomic units)

$$\left(-\frac{1}{2}\nabla^2 + U(\mathbf{r}) + V_H(\mathbf{r}) + V_{xc}\mathbf{r})\right)\psi_i(\mathbf{r}) = \varepsilon_i\psi_i(\mathbf{r}), \tag{5.24}$$

where

$$V_H(\mathbf{r}) = \int d\mathbf{r}' \frac{n(\mathbf{r}')}{|\mathbf{r}-\mathbf{r}'|} \tag{5.25}$$

is the Hartree potential, and

$$V_{xc}(\mathbf{r}) = \frac{\delta E_{xc}[n]}{\delta n} \tag{5.26}$$

is the exchange-correlation potential. The density can then be calculated as

$$n(\mathbf{r}) = \sum_{i}^{N} |\psi_i(\mathbf{r}')|^2, \tag{5.27}$$

which is in principle exact. The difficulty with the density functional theory is that the exchange-correlation functional E_{xc} is not known and can be only approximately obtained for the homogeneous electron gas. A common approximation is the local density approximation (LDA) where the non-uniform electron gas is treated as if it was a uniform electron gas at constant density n at a given position. An improvement of this approach for inhomogeneous systems is called the generalized gradient approximation where the exchange-correlation energy is expressed as a non-linear function of local density and its gradient.

We now discuss the application of DFT to graphene quantum dots [33, 44]. In Fig. 5.2, we show the results of the density functional calculation within the local density approximation using SIESTA for a 97 carbon atom triangular graphene quantum dot. The zigzag edges are passivated with hydrogen atoms, as already discussed in Sect. 4.1 (also see Fig. 4.3d). The quasiparticle energy levels obtained from DFT are compared to tight-binding (TB) and Hartree-Fock (TB+HF) calculations. In Hartree-Fock and DFT calculations, seven electrons were removed in order to leave the zero-energy states empty. The effect of having electrons in the zero-energy shell will be

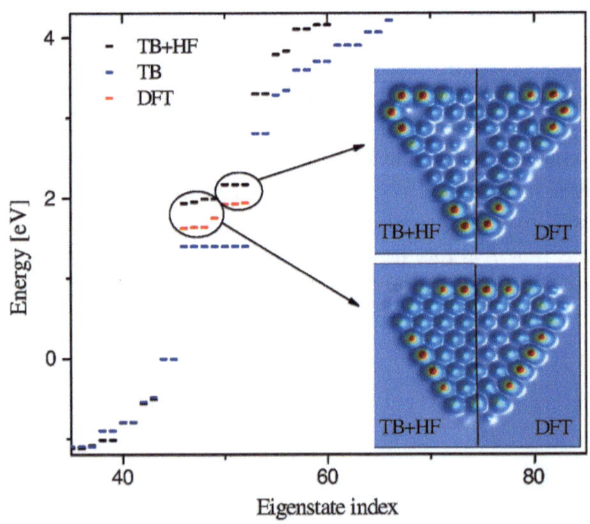

Fig. 5.2 Single particle spectrum of a triangular graphene island of 97 carbon atoms obtained by tight-binding (TB, blue lines) and self-consistent Hartree-Fock (TB+HF, black lines) methods. The 7 zero-energy states near the Fermi level are compared to DFT results. In Hartree-Fock and DFT calculations 7 electrons were removed, leaving the zero-energy states empty. The dielectric constant κ is set to 6. Inset compares the structure of corner and side states obtained using Hartree-Fock and DFT calculations. In DFT calculations, hydrogen atoms were attached to dangling bonds. Reprinted from [33]

studied in detail in Chap. 6. As was explained in Sect. 4.1, the zigzag edges lead to a band of degenerate levels in the TB calculations. However, due to the mean-field interaction with the valence electrons, a group of three states is now separated from the rest by a small gap of ~0.2 eV. The three states correspond to quasiparticles localized in the three corners of the triangle. The same physics occurs both in TB+HF and DFT calculations. In the inset of Fig. 5.2, we also compare the electronic densities, showing that TB+HF and DFT results are in good agreement [33].

5.6 Configuration Interaction Method

The Configuration Interaction (CI) method is the most direct and accurate way of solving the many-body Schrödinger equation. Although computationally very demanding, it captures all correlation effects missing in DFT and HF calculations.

We start by writing the full many-body Hamiltonian of interacting electrons occupying single-particle energy levels as

$$H_{MB} = \sum_{s,\sigma} E_{s\sigma} a^{\dagger}_{s\sigma} a_{s\sigma} + \frac{1}{2} \sum_{\substack{s,p,d,f,\\ \sigma,\sigma'}} \langle sp \mid V \mid df \rangle a^{\dagger}_{s\sigma} a^{\dagger}_{p\sigma'} a_{d\sigma'} a_{f\sigma}. \tag{5.28}$$

In the first term, the energies $E_{s\sigma}$ correspond to eigenvalues of TB Hamiltonian given by (5.14). The second term describes scattering between pairs of quasi-particles from energy levels d, f to s, p. The two-body quasi-particle scattering matrix elements $\langle sp \mid V \mid df \rangle$ are calculated from the two-body localized on-site Coulomb matrix elements $\langle ij \mid V \mid kl \rangle$. Because the Hamiltonian given by (5.28) does not contain any spin interaction terms, total spin S and its projection onto z axis, S_z, are good quantum numbers. The Hamiltonian matrix can be divided into blocks corresponding to different S or S_z. Each block can be diagonalized independently. In the next subsections we discuss the method of constructing the many-body basis of configurations for a given S_z and constructing the Hamiltonian matrix in the space of configurations.

5.6.1 Many-Body Configurations

For a given number of electrons we write a many-body configuration consisting of electrons distributed on single-particle orbitals, written as

$$|\Psi_1\rangle = \prod_{s\sigma} a^{\dagger}_{s\sigma} |0\rangle,$$

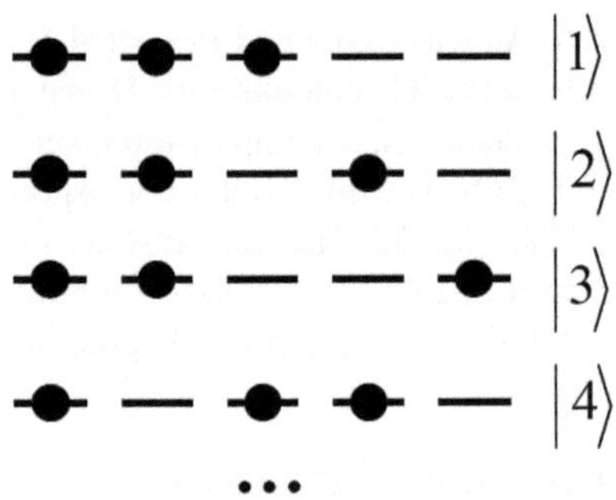

Fig. 5.3 A scheme of possible distributions of spinless $N_{el} = 3$ particles within $N_{st} = 5$ energy states. *Black* bars correspond to energy levels and black circles to electrons. One can construct the total of $N_{conf} = 10$ distinct configurations. They form a many-body basis in the configuration interaction method

where $|0\rangle$ is the vacuum state. The number of operators in this product is equal to the number of electrons N_{el}. We now show how to construct a complete set of basis vectors on an example of $N_{el} = 3$ particles distributed within $N_{st} = 5$ states, for simplicity neglecting spin degrees of freedom. This is schematically presented in Fig. 5.3. Black bars corresponds to energy levels and black circles to electrons. The first configuration from Fig. 5.3 can be written as

$$|1\rangle = a_1^\dagger a_2^\dagger a_3^\dagger |0\rangle,$$

where numbers $1, 2\ldots$ label energy levels counted from the left to the right. We note that in order to avoid double counting of the same configuration one has to choose some convention of ordering creation operators in the many-body vectors. Our choice is that we always write $...a_i^\dagger a_j^\dagger ...|0>$ for $i < j$. The total number of possible configurations N_{conf} is given by $N_{conf} = \begin{pmatrix} N_{st} \\ N_{el} \end{pmatrix}$. Thus, one can construct N_{conf} linearly independent vectors which span the Hilbert space. For the example in Fig. 5.3, $N_{st} = 5$ and $N_{el} = 3$, and $N_{conf} = 10$.

We now include spin degrees of freedom. A many-body configuration is a product of configurations for spin-down and spin-up configurations

$$|\Psi_1\rangle = \prod_s a_{s\downarrow}^\dagger |0\rangle \otimes \prod_p a_{p\uparrow}^\dagger |0\rangle = ...a_{s\downarrow}^\dagger ...a_{p\uparrow}^\dagger ...|0\rangle, \tag{5.29}$$

with the number of creation operators in this product equal to the number of electrons $N_{el} = N_{dn} + N_{up}$, and N_{dn} (N_{up}) defined as the number of electrons with spin down (spin up). The total number of configurations is $N_{conf} = N_{conf}^{dn} \cdot N_{conf}^{up}$. The operator corresponding to the projection of total spin S onto z-axis is defined as $\hat{S}_z = \sum_s \sigma a_{s\sigma}^\dagger a_{s\sigma}$. This operator commutes with the Hamiltonian given by (5.28), $[\hat{H}, \hat{S}_z] = 0$. Additionally, many-body configurations given by (5.29) are eigenvectors of this operator: each of them has a well defined projection of spin onto z-axis, $S_z = (N_{up} - N_{dn})/2$. For a given number of particles, N_{el}, sets of vectors for each S_z are constructed. These vectors span independent subspaces and the Hamiltonian matrix can be written in a block diagonal form. One can also block diagonalize the Hamiltonian matrix using eigenstates of the total spin $\hat{S}^2 = \frac{N_{el}}{2} + \hat{S}_z^2 - \sum_{sp} a^\dagger_{s\uparrow} a^\dagger_{p\downarrow} a_{p\downarrow} a_{s\uparrow}$ operator, which commutes with the Hamiltonian $[\hat{H}, \hat{S}^2] = 0$ [45]. However, the cost of the rotation of the basis into the eigenstates of the total spin often outweighs the benefits. Therefore, we usually calculate the eigenstates of $\hat{S}_z$ and deduce the eigenstates of $\hat{S}^2$ [45].

Once the many-body basis set is constructed, we can then proceed with the construction of the Hamiltonian given by (5.28). The main difficulty is the calculation of matrix elements of the Coulomb interaction term $\hat{V}$. For completeness we briefly outline here how this is done as results are discussed throughout this monograph.

As an example let us choose a system with 4 electrons with $S_z = 0$, two spin-down and two spin-up, distributed on 3 single-particle states for each spin. There are $N_{conf} = 3 \cdot 3 = 9$ possible configurations:

$$\begin{aligned}
|\Psi_1\rangle &= a^\dagger_{1\downarrow} a^\dagger_{2\downarrow} a^\dagger_{1\uparrow} a^\dagger_{2\uparrow} |0\rangle, & |\Psi_6\rangle &= a^\dagger_{1\downarrow} a^\dagger_{3\downarrow} a^\dagger_{2\uparrow} a^\dagger_{3\uparrow} |0\rangle,\\
|\Psi_2\rangle &= a^\dagger_{1\downarrow} a^\dagger_{2\downarrow} a^\dagger_{1\uparrow} a^\dagger_{3\uparrow} |0\rangle, & |\Psi_7\rangle &= a^\dagger_{2\downarrow} a^\dagger_{3\downarrow} a^\dagger_{1\uparrow} a^\dagger_{2\uparrow} |0\rangle,\\
|\Psi_3\rangle &= a^\dagger_{1\downarrow} a^\dagger_{2\downarrow} a^\dagger_{2\uparrow} a^\dagger_{3\uparrow} |0\rangle, & |\Psi_8\rangle &= a^\dagger_{2\downarrow} a^\dagger_{3\downarrow} a^\dagger_{1\uparrow} a^\dagger_{3\uparrow} |0\rangle,\\
|\Psi_4\rangle &= a^\dagger_{1\downarrow} a^\dagger_{3\downarrow} a^\dagger_{1\uparrow} a^\dagger_{2\uparrow} |0\rangle, & |\Psi_9\rangle &= a^\dagger_{2\downarrow} a^\dagger_{3\downarrow} a^\dagger_{2\uparrow} a^\dagger_{3\uparrow} |0\rangle.\\
|\Psi_5\rangle &= a^\dagger_{1\downarrow} a^\dagger_{3\downarrow} a^\dagger_{1\uparrow} a^\dagger_{3\uparrow} |0\rangle, & &
\end{aligned} \tag{5.30}$$

We keep the convention that from the left to the right we have creation operators with increasing indices, first for spin-down, and next for spin-up operators.

We now rewrite the two-body scattering term $\hat{V} = \frac{1}{2} \sum_{s,p,d,f} \langle sp \mid V \mid df\rangle a^\dagger_s a^\dagger_p a_d a_f$ in a convenient form, with combined orbital and spin quantum numbers, e.g., $s \equiv s\sigma$, $p \equiv p\sigma'$, etc. In the next step we divide the summation over the indices of the initial d, f and final s, p pairs into two parts, for $d < f$ and $d > f$ and $s > p$ and $s < p$. This removes the factor of $1/2$ and introduces explicitly the direct and exchange scattering matrix elements. The final form of the two-body Coulomb operator is

$$\hat{V} = \sum_{s>p,d<f} (\langle sp \mid V \mid df\rangle - \langle sp \mid V \mid fd\rangle) \, a^\dagger_s a^\dagger_p a_d a_f. \tag{5.31}$$

We can now bring back the explicit spin dependence and write the two-body scattering term V as a sum of different spin contributions, $\hat{V} = \hat{V}_{\downarrow\downarrow} + \hat{V}_{\uparrow\uparrow} + \hat{V}_{\downarrow\uparrow}$,.

In order to construct the matrix element of $\hat{V}$ we show how it acts on one of the basis configurations $|\Psi_8\rangle$:

$$\hat{V}_{\downarrow\downarrow} |\Psi_8\rangle \to \sum_{s>p,d<f} a^\dagger_{s\downarrow} a^\dagger_{p\downarrow} a_{d\downarrow} a_{f\downarrow} a^\dagger_{2\downarrow} a^\dagger_{3\downarrow} a^\dagger_{1\uparrow} a^\dagger_{3\uparrow} |0\rangle. \tag{5.32}$$

There are two spin-down annihilation operators and two spin-down creation operators. They act only on spin-down electrons, removing the pair 2, $\downarrow$, 3, $\downarrow$, and replacing it with a new pair s, $\downarrow$, p, $\downarrow$:

$$\hat{V}_{\downarrow\downarrow}|\Psi_8\rangle \to \sum_{s>p} a^{\dagger}_{s\downarrow} a^{\dagger}_{p\downarrow} a_{2\downarrow} a_{3\downarrow} a^{\dagger}_{2\downarrow} a^{\dagger}_{3\downarrow} a^{\dagger}_{1\uparrow} a^{\dagger}_{3\uparrow}|0\rangle. \tag{5.33}$$

All possible new four-electron states are given by

$$\hat{V}_{\downarrow\downarrow}|\Psi_8\rangle \to -\sum_{s>p} a^{\dagger}_{s\downarrow} a^{\dagger}_{p\downarrow} a^{\dagger}_{1\uparrow} a^{\dagger}_{3\uparrow}|0\rangle. \tag{5.34}$$

The electron operators in the final four electron states have to be reordered to conform to the original choice of the basis configuration,

$$\hat{V}_{\downarrow\downarrow}C = a^{\dagger}_{1\downarrow} a^{\dagger}_{2\downarrow} a^{\dagger}_{1\uparrow} a^{\dagger}_{3\uparrow}|0\rangle + a^{\dagger}_{1\downarrow} a^{\dagger}_{3\downarrow} a^{\dagger}_{1\uparrow} a^{\dagger}_{3\uparrow}|0\rangle + a^{\dagger}_{2\downarrow} a^{\dagger}_{3\downarrow} a^{\dagger}_{1\uparrow} a^{\dagger}_{3\uparrow}|0\rangle = |\Psi_2\rangle + |\Psi_5\rangle + |\Psi_8\rangle.$$

Hence, the two body term $\hat{V}_{\downarrow\downarrow}$ acting on $|\Psi_8\rangle$ created also $|\Psi_2\rangle$ and $|\Psi_5\rangle$ with amplitude given by, e.g., $\langle\Psi_2|\hat{V}_{\downarrow\downarrow}|\Psi_8\rangle = \langle 21 \mid V \mid 23\rangle - \langle 21 \mid V \mid 32\rangle$. In a similar but much more efficient way all matrix elements $\langle i|H|j\rangle$ of the Hamiltonian are constructed in the space of configurations $|i\rangle$.

5.6.2 Diagonalization Methods for Large Matrices

In the configuration interaction method, the size of the Hilbert space increases exponentially with the number of particles and the number of states. For example, for a system with $N_{st} = 10$ and $N_{el} = 5$ one gets

$$N_{conf} = \binom{10}{5} = \frac{10!}{(10-5)!5!} = 126 \simeq 10^2,$$

but if one doubles the number of states and particles,

$$N_{conf} = \binom{20}{10} = \frac{20!}{(20-10)!10!} = 184756 \simeq 10^5,$$

which is three orders of magnitude larger. Thus, this method is restricted to calculations of small systems and an efficient computational methods is required.

For large matrices, $N_{conf} > 10^5$, it is difficult to store all matrix elements due to their large numbers. Moreover, standard diagonalization procedures used in linear algebra packages, e.g. Lapack subroutines, become extremely costly in terms of computation time. However, from the physical point of view, one usually needs only a few lowest eigenvalues or eigenstates of the Hamiltonian that correspond to the ground state and low-energy excited states. In order to find these eigenstates, iterative methods, such as Lanczos, are required. The Lanczos method allows to find extremal eigenvalues of large matrices [46]. The Lanczos method is based on the matrix-vector multiplication and in each consecutive iteration only the product of this operation is required. The efficient way to overcome problems with storing matrix elements is to calculate them "on the fly", separately for each iteration. Calculated matrix elements are multiplied by appropriate coefficients of a given vector and only a product of this operation, a vector, is stored. Thus, instead of storing N_{conf}^2 matrix elements, one can only store N_{conf} coefficients of the new vector.

5.7 TB+HF+CI Method

We now turn to study the role of electron–electron interactions in graphene nanostructures. Solving the full many-body problem, even for structures with tens of atoms, is not possible at present. Thus, we combine the tight-binding approach with the mean-field HF method and with the exact CI diagonalization method. We are interested in quantum dots with degenerate energy shells, where electron–electron interactions play a critically important role. Thus, we explain our methodology by applying it to a small TGQD consisting of $N = 97$ atoms with $N_{deg} = 7$ degenerate states [33]. The procedure is schematically shown in Fig. 5.4. In Fig. 5.4a we clearly see that the valence band and the degenerate shell are separated by the energy gap. Thus, the closed-shell system of $N_{ref} = N - N_{deg}$ interacting electrons is expected to be well described in a mean-field approximation, using a single Slater determinant. This corresponds to a charged system with N_{deg} positive charges, as schematically shown in Fig. 5.4b. The Hamiltonian given by (5.20) is self-consistently solved for N_{ref} electrons. New orbitals obtained for quasi-particles correspond to a fully occupied valence band and completely empty degenerate states. One can note that because of the mean-field interaction with the valence electrons, a group of three states is now separated from the rest by a small gap of ~0.2 eV, Fig. 5.4b. The three states correspond to HF quasiparticles localized in the three corners of the triangle [33]. As will be shown in Sect. 4.1.3, this is related to the long-range Coulomb interaction. We start filling the degenerate energy levels by adding extra electrons one by one, as schematically shown in Fig. 5.4c. Next, we solve the many-body Hamiltonian corresponding to the added electrons, given by (5.28). In our calculations, we neglect scattering from/to the states from a fully occupied valence band. Moreover, because of the large energy gap between the degenerate states and the conduction band, we can neglect scatterings to the higher energy states. Our assumptions can be confirmed by comparing the energy gaps and Coulomb interaction matrix elements. For example, the system with degenerate states separated by energy gaps $\Delta E \sim 0.5$ eV has the intra-degenerate states interaction terms $V \sim 0.23$ eV. The Coulomb matrix elements V scattering electrons from an arbitrary degenerate state to the valence band and/or to the conduction band are $V \sim 0.2$ eV. Hence, the effect of these scattering processes

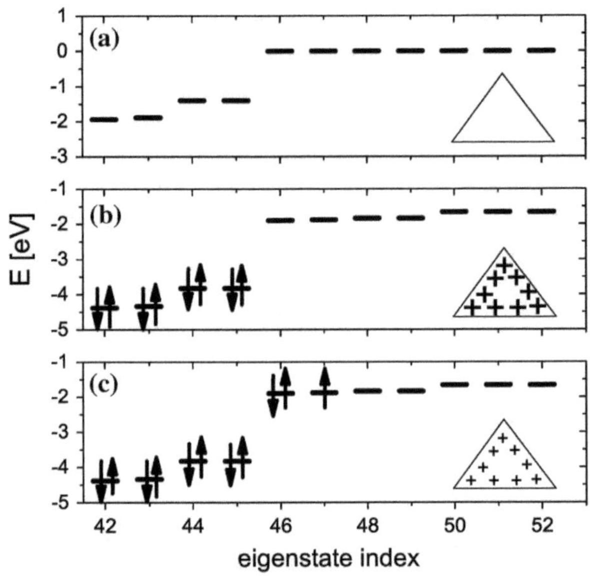

Fig. 5.4 **a** Single-particle nearest-neighbor TB energy levels. The zero-energy shell on the Fermi level is perfectly degenerate. **b** Positively charged system with an empty degenerate band after self-consistent Hartree-Fock (HF) mean-field calculations described by a single Slater determinant (TB+HF model). **c** Occupation of empty degenerate HF quasi-orbitals by electrons. The inset pictures schematically show the excess charge corresponding to each of the three model systems. The ground state and the total spin of the system of interacting electrons can be calculated by using the configuration interaction (CI) method. The charge neutrality corresponds to a half-filled degenerate band (not shown)

is proportional to $V^2/\Delta E = \delta$, where $\delta \ll \Delta E$ and the effect is weak. Thus, many-body properties of electrons occupying the degenerate states are primarily governed by interactions between electrons within these states. These approximations allow us to treat the degenerate shell as an independent system which significantly reduces the dimension of the Hilbert space. The basis is constructed from vectors corresponding to all possible many-body configurations of electrons distributed within the degenerate states. For a given number of electrons, N_{el}, the Hamiltonian given by (5.28) is diagonalized in each subspaces with a given S_z. The results of the TB+HF+CI method applied to graphene quantum dots are discussed in the following chapters.

References

1. G.D. Mahan, *Many-Particle Physics* (Plenum Press, New York, 1990)
2. J.J. Quinn, K.-S. Yi, *Solid State Physics: Principles and Modern Applications* (Springer, Heidelberg, 2009)

3. G. Giuliani, G. Vignale, *Quantum Theory of the Electron Liquid* (Cambridge University Press, Cambridge, 2008)
4. P. Hawrylak, Phys. Rev. Lett. **59**, 485 (1987)
5. P. Hawrylak, S. Das Sarma, Advances in studies of electrons in low dimensional structures. Solid State Commun. **127**, 753 (2003)
6. P. Hawrylak, M. Korkusinski, Electronic and optical properties of self-assembled quantum dots, in *Single Quantum Dots: Fundamentals, Applications, and New Concepts*, vol. 90, Topics in Applied Physics, ed. by P. Michler (Springer, Heidelberg, 2003), pp. 25–92
7. C.-Y. Hsieh, Y.P. Shim, M. Korkusinski, P. Hawrylak, Physics of triple quantum dot molecule with controlled electron numbers. Rep. Prog. Phys. **75**, 114501 (2012)
8. G. Dresselhaus, M.S. Dresselhaus, Adv. Phys. **30**(2), 139–326 (1981)
9. K. Shung, Phys. Rev. B **34**, 979993 (1986)
10. P. Hawrylak, Solid State Commun. **63**, 241 (1987)
11. A.H. Castro Neto, F. Guinea, N.M.R. Peres, K.S. Novoselov, A.K. Geim, Rev. Mod. Phys. **81**, 109 (2009)
12. S. Das Sarma, S. Adam, E.H. Hwang, E. Rossi, Rev. Mod. Phys. **83**, 407 (2011)
13. V.N. Kotov, B. Uchoa, V.M. Pereira, F. Guinea, A.H. Castro Neto, Rev. Mod. Phys. **84**, 10671125 (2012)
14. M.A.H. Vozmediano, F. Guinea, Phys. Scr. **T146**, 014015 (2012)
15. M.I. Katsnelson, *Graphene: Carbon in Two Dimensions* (Cambridge University Press, Cambridge, 2012)
16. A.H. MacDonald, J. Jung, F. Zhang, Phys. Scr. **T146**, 014012 (2012)
17. A. Hideo, D. Mildred (eds.), *Physics of Graphene* (Springer, Heidelberg, 2014)
18. L.E.F. Foa Torres, S. Roche, J.C. Charlier, *Introduction to Graphene Based Nanomaterials: From Electronic Structure to Quantum Transport* (Cambridge University Press, Cambridge, 2014)
19. C. Wang, D. Wong, A.V. Shytov, V.W. Brar, S. Choi, Q. Wu, H.-Z. Tsai, W. Regan, A. Zettl, R.K. Kawakami, S.G. Louie, L.S. Levitov, M.F. Crommie, Science **340**, 734 (2013)
20. C. Toke, P.E. Lammert, V.H. Crespi, J.K. Jain, Phys. Rev. B **74**, 235417 (2006)
21. V.M. Apalkov, T. Chakraborty, Phys. Rev. Lett. **97**, 126801 (2006)
22. N. Shibata, K. Nomura, J. Phys. Soc. Jpn. **78**, 104708 (2009)
23. L. Yang, M.L. Cohen, S.G. Louie, Nano Lett. **7**, 3112 (2007)
24. T. Wassmann, A.P. Seitsonen, A.M. Saitta, M. Lazzeri, F. Mauri, Phys. Rev. Lett. **101**, 096402 (2008)
25. P. Koskinen, S. Malola, H. Häkkinen, Phys. Rev. Lett. **101**, 115502 (2008)
26. K.T. Chan, J.B. Neaton, M.L. Cohen, Phys. Rev. B **77**, 235430 (2008)
27. H. Sevincli, M. Topsakal, E. Durgun, S. Ciraci, Phys. Rev. B **77**, 195434 (2008)
28. I. Ozfidan, M. Korkusinski, A.D. Güçlü, J. McGuire P. Hawrylak, Phys. Rev. B **89**, 085310 (2014)
29. S. Sorella, E. Tosatti, Europhys. Lett. **19**, 699 (1992)
30. T.O. Wehling, E. Sasioglu, C. Friedrich, A.I. Lichtenstein, M.I. Katsnelson, S. Blügel, Phys. Rev. Lett. **106**, 236805 (2011)
31. E.K.U. Gross, E. Runge, O. Heinonen, *Many-Particle Physics* (Adam Hilger, New York, 1991)
32. B.J. Ransil, Rev. Mod. Phys. **32**, 245 (1960)
33. A.D. Güçlü, P. Potasz, O. Voznyy, M. Korkusinski, P. Hawrylak, Phys. Rev. Lett. **103**, 246805 (2009)
34. P. Potasz, A.D. Güçlü, P. Hawrylak, Phys. Rev. B **82**, 075425 (2010)
35. P.R. Wallace, Phys. Rev. **71**, 622 (1947)
36. J. Hubbard, Proc. R. Soc. Lond. Ser. A **276**, 238 (1963)
37. K.A. Chao, J. Spałek, A.M. Oleś, J. Phys. C: Solid State Phys. **10**, L271 (1977)
38. K.A. Chao, J. Spałek, A.M. Oleś, Phys. Rev. B **18**, 3453 (1978)
39. M. Ezawa, Phys. Rev. B **76**, 245415 (2007)
40. J. Fernandez-Rossier, J.J. Palacios, Phys. Rev. Lett. **99**, 177204 (2007)
41. O.V. Yazyev, Phys. Rev. Lett. **101**, 037203 (2008)

42. P. Hohenberg, W. Kohn, Phys. Rev. **136**, B864 (1964)
43. W. Kohn, L.J. Sham, Phys. Rev. **140**, A1133–A1138 (1965)
44. O. Voznyy, A.D. Güçlü, P. Potasz, P. Hawrylak, Phys. Rev. B **83**, 165417 (2011)
45. A. Wensauer, M. Korkusiński, P. Hawrylak, Solid State Commun. **130**, 115 (2004)
46. C. Lanczos, J. Natl. Bureau Standards **45**, 255 (1950)

Chapter 6
Magnetic Properties of Gated Graphene Nanostructures

Abstract In this chapter we describe magnetic properties of graphene quantum dots and rings with broken sublattice symmetry using the TB+HF+CI methodology. The broken sublattice symmetry leads to the existence of a shell of degenerate levels at the Fermi level. We discuss how the electronic and magnetic properties of GQDs depend on the filling of the shell in triangular graphene quantum dots (TGQD), how they can be controlled by electric field in bi-layer TGQDs and how they can be detected in Coulomb and Spin Blockade transport experiments.

6.1 Triangular Graphene Quantum Dots with Zigzag Edges

Here, we discuss the magnetic properties of triangular graphene quantum dots (TGQD) with zigzag edges. A theorem due to Lieb allows prediction of magnetic properties of charge-neutral graphene quantum dots within the Hubbard model [1]. We find that while the Lieb's theorem holds for charge-neutral quantum dots even beyond the Hubbard model, as we add or remove electrons from it, electronic correlations play a dominant role in determining magnetic, electronic, transport, and optical properties [2, 3].

6.1.1 Filling Factor Dependence of the Total Spin of TGQD

In Fig. 6.1 we analyze the dependence of the low-energy spectra obtained using TB+HF+CI methodology on the total spin S for a charge-neutral TGQD with $N=97$ carbon atoms and $N_{el}=7$ electrons on the degenerate shell with seven states (Fig. 6.1a), and charged TGQD, i.e., $N_{el}=8$ electrons (Fig. 6.1b). We see that for the charge-neutral TGQD with $N_{el}=7$ electrons the ground state is maximally spin polarized, with $S=3.5$, indicated by a circle. There is only one possible configuration of all electrons with parallel spins that corresponds to exactly one electron per one degenerate state. The energy of this configuration is well separated from other states with lower total spin S, which requires at least one flipped spin among seven initially

A.D. Güçlü et al., *Graphene Quantum Dots*,
NanoScience and Technology, DOI 10.1007/978-3-662-44611-9_6

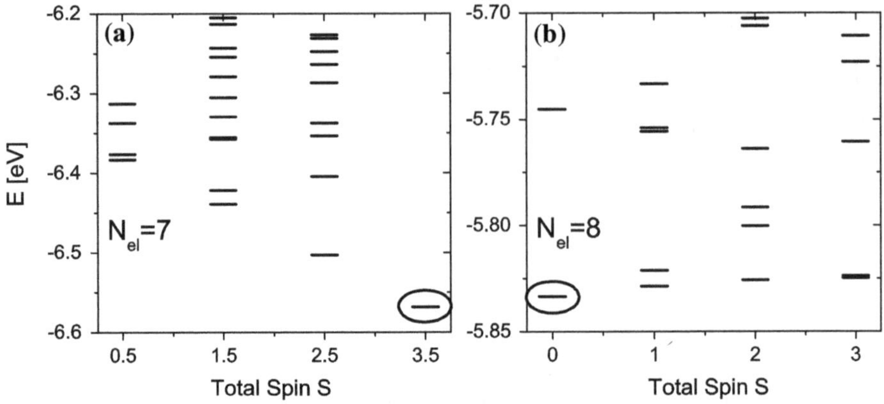

Fig. 6.1 The low-energy spectra for the different total spin S for **a** $N_{el} = 7$ electrons and **b** $N_{el} = 8$ electrons. For $N_{el} = 7$ electrons the ground state corresponding to $S = 3.5$, indicated by a *circle*, is well separated from excited states with different total spin S. For $N_{el} = 8$ electrons the ground state corresponding to $S = 0$, indicated by a *circle*, is almost degenerate with excited states with different total spin S. Reprinted from [3]

spin-polarized electrons. When an extra electron is added through, for instance, an external gate, the spectrum changes drastically as seen in Fig. 6.1b. In particular, the ground state is now depolarized with $S = 0$, indicated by a circle. This new ground state is almost degenerate with states corresponding to the different total spin, which is a signature of strong electronic correlations.

The calculated many-body energy levels, including all spin states for different numbers of electrons (shell filling), are shown in Fig. 6.2. For each electron number, N_{el}, energies are measured from the ground-state energy and scaled by the energy gap of the half-filled shell, corresponding to $N_{el} = 7$ electrons in this case. The solid line shows the evolution of the energy gap as a function of shell filling. The energy gaps for a neutral system, $N_{el} = 7$, as well as for $N_{el} = 7-3 = 4$ and $N_{el} = 7+3 = 10$ are found to be significantly larger in comparison to the energy gaps for other electron numbers. In addition, close to the half-filled degenerate shell, the reduction of the energy gap is accompanied by an increase of low-energy density of states indicating strong electronic correlations. These results show that correlations effects can play an important role at different filling factors.

We now extract the total spin and energy gap for each electron number. Figures 6.3a, b show the phase diagram, the total spin S and an excitation gap, as a function of the number of electrons occupying the shell. The TGQD reveals maximum spin polarization for almost all electron numbers, with exceptions for $N_{el} = 8, 9$ electrons. However, the energy gaps are found to strongly oscillate as a function of shell filling as a result of a combined effect of correlations and system's geometry. We observe a competition between fully spin polarized TGQD that maximizes exchange energy and fully unpolarized system that maximizes the correlation energy. Only close to the charge neutrality, for $N_{el} = 8$ and $N_{el} = 9$ electrons, are

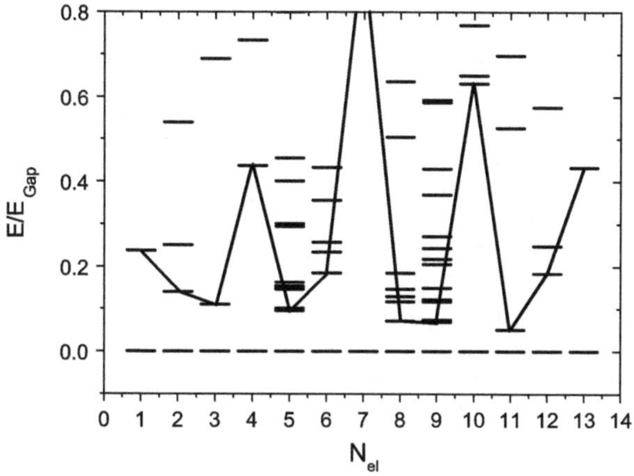

Fig. 6.2 The low-energy spectra of the many-body states as a function of the number of electrons occupying the degenerate shell for the system with $N_{deg} = 7$ degenerate states. The energies are renormalized by the energy gap corresponding to the half-filled shell, $N_{el} = 7$ electrons. The *solid line* shows the evolution of the energy gap as a function of shell filling. Reprinted from [3]

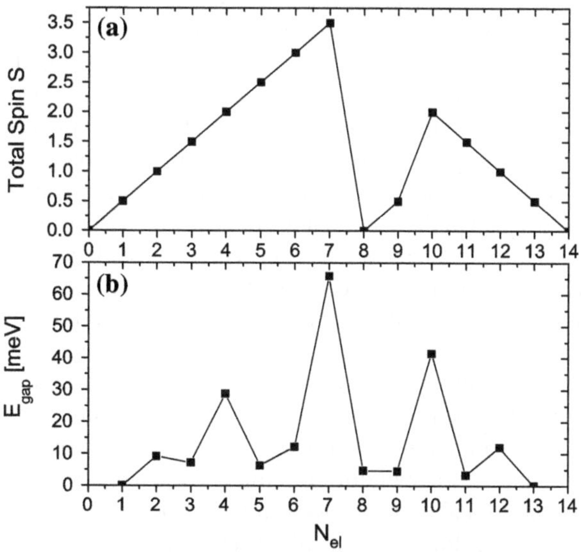

Fig. 6.3 **a** The total spin as a function of the number of electrons occupying the degenerate shell and **b** corresponding the energy excitation gaps. Due to a presence of correlation effects for some fillings, the magnitude of the energy gap is significantly reduced. Reprinted from [3]

the correlations sufficiently strong to overcome the large cost of the exchange energy related to flipping spin. The excitation gap is significantly reduced and exhibits large density of states at low energies, as shown in Fig. 6.1. Away from half-filling, we observe larger excitation gaps for $N_{el} = 4$ and $N_{el} = 10$ electrons. These fillings

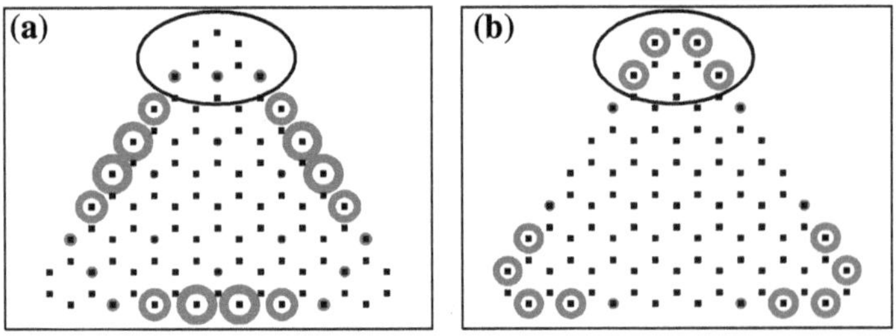

Fig. 6.4 The spin densities of the ground state for **a** $N_{el} = 4$ electrons and **b** $N_{el} = 10$ electrons that correspond to subtracting/adding three electrons from/to the charge-neutral system. The radius of *circles* is proportional to a value of spin density on a given atom. A long range Coulomb interaction repels **a** holes and **b** electrons to three corners, forming a spin-polarized Wigner-like molecule. Reprinted from [3]

correspond to subtracting/adding three electrons from/to the charge-neutral system with $N_{el} = 7$ electrons. In Fig. 6.4 we show the corresponding spin densities. Here, long range interactions dominate the physics and three spin polarized (Fig. 6.4a) holes ($N_{el} = 7 - 3$ electrons) and (Fig. 6.4b) electrons ($N_{el} = 7 + 3$ electrons) maximize their relative distance by occupying three consecutive corners. Electron spin density is localized in each corner while holes correspond to missing spin density localized in each corner. We also note that this is not observed for $N_{el} = 3$ electrons filling the degenerate shell (not shown here). The energies of HF orbitals of corner states correspond to three higher energy levels (see Fig. 5.4c), with electronic densities shown in [2]. Thus, $N_{el} = 3$ electrons occupy lower-energy degenerate levels corresponding to sides instead of corners. On the other hand, when $N_{el} = 7$ electrons are added to the shell, the HF quasiparticle energies are renormalized. The degenerate shell is again almost perfectly flat similarly to levels obtained within the TB model. The kinetic energy does not play any role allowing a formation of a spin-polarized Wigner-like molecule, resulting from long-range interactions and a triangular geometry. We note that Wigner molecules were previously discussed in circular graphene quantum dots with zigzag edges described in the effective mass approximation [4, 5]. The rotational symmetry of the quantum dot allowed for the construction of an approximate correlated ground state corresponding to either a Wigner crystal or Laughlin-like state [4]. Later, a variational rotating-electron-molecule (VREM) wave function was used [5]. Unfortunately, due to a lack of an analytical form of a correlated wave function with a triangular symmetry, it is not possible to do it here.

6.1.2 Size Dependence of Magnetic Properties of TGQD: Excitons, Trions and Lieb's Theorem

In the previous section we have analyzed in detail the electronic properties of a particular TGQD with $N = 97$ atoms as a function of the filling factor $\nu = N_{el}/N_{deg}$,

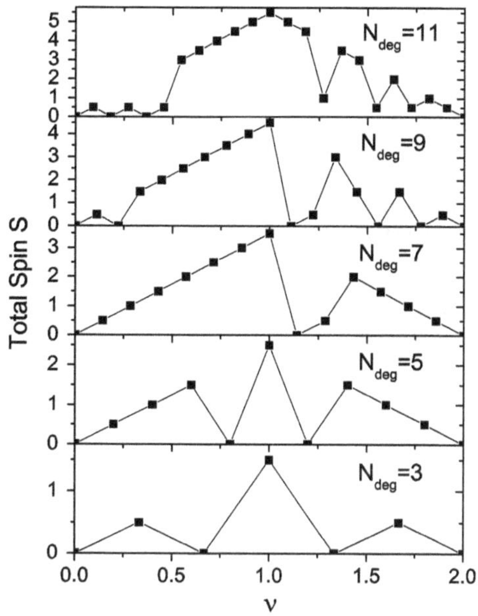

Fig. 6.5 Spin phase diagrams as a function of filling factor $\nu = N_{el}/N_{deg}$ for different size *triangles* characterized by the number of the degenerate edge states N_{deg}. Half-filled shell $\nu = 1$ is always maximally spin polarized. The complete spin depolarization occurs for one added electron to the charge-neutral system for $N_{deg} \leq 9$. For $N_{deg} = 11$ the depolarization effect moves to a different filling. Reprinted from [3]

i.e., the number of electrons per the number of degenerate levels. In this section we address the important question of whether one can predict the electronic properties of a TGQD as a function of size.

Figure 6.5 shows spin phase diagrams for triangles with odd number of degenerate edge states N_{deg} and increasing size. Clearly, the total spin depends on the filling factor and size of the triangle. However, all charge-neutral systems at $\nu = 1$ are always maximally spin-polarized and a complete depolarization occurs for $N_{deg} \leq 9$ for structures with one extra electron added (such depolarization also occurs for even N_{deg}, not shown). However, at $N_{deg} = 11$ we do not observe depolarization for $N_{deg} + 1$ electrons but for $N_{deg} + 3$, where a formation of Wigner-like molecule for a triangle with $N_{deg} = 7$ was observed. We will come back to this problem later. We now focus on the properties close to the charge neutrality.

For the charge-neutral case, the ground state corresponds to only one configuration $|GS\rangle = \prod_i a^\dagger_{i,\downarrow}|0\rangle$ with maximum total S_z and occupation of all degenerate shell levels i by electrons with parallel spin. Here $|0\rangle$ is the HF ground state of all valence electrons. Let us consider the stability of the spin polarized state to single spin flips. We construct spin-flip excitations $|kl\rangle = a^\dagger_{k,\uparrow}a_{l,\downarrow}|GS\rangle$ from the spin-polarized degenerate shell. The spin-up electron interacts with a spin-down "hole" in a spin-polarized state and forms a collective excitation, an exciton. An exciton spectrum is obtained by building an exciton Hamiltonian in the space of electron-hole pair excitations and diagonalizing it numerically, as was done, e.g., for semiconductor quantum dots [6]. If the energy of the spin flip excitation turns out to be negative in comparison with the spin-polarized ground state, the exciton is bound

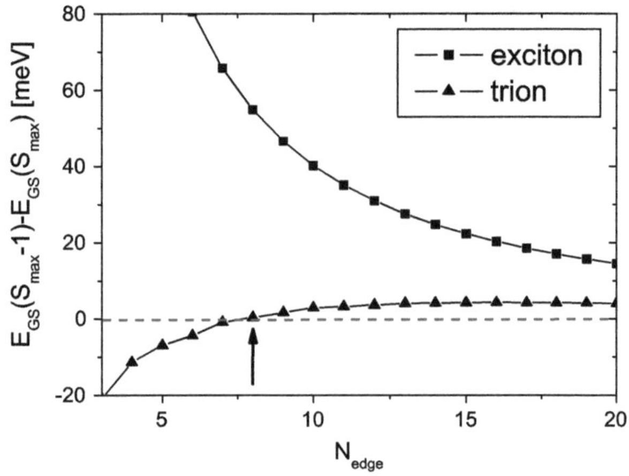

Fig. 6.6 Size-dependent analysis based on exciton and trion binding energies. For the charge-neutral system, it is energetically unfavorable to form an exciton, which is characterized by a positive binding energy. The formation of a trion is desirable for small size systems. The phase transition occurs close to $N_{deg} = 8$, indicated by an *arrow*. Reprinted from [3]

and the spin-polarized state is unstable. The binding energy of a spin-flip exciton is a difference between the energy of the lowest state with $S = S_z^{max} - 1$ and the energy of the spin-polarized ground state with $S = S_z^{max}$. An advantage of this approach is the ability to test the stability of the spin polarized ground state for much larger TGQD sizes.

Figure 6.6 shows the exciton binding energy as a function of the size of TGQD, labeled by a number of the degenerate states N_{deg}. The largest system, with $N_{deg} = 20$, corresponds to a structure consisting of $N = 526$ atoms. The exciton binding energies are always positive, i.e., the exciton does not form a bound state, confirming a stable magnetization of the charge neutral system. The observed ferromagnetic order was also found by other groups based on calculations for small systems with different levels of approximations [2, 7–9]. The above results confirm predictions based on Lieb's theorem for the Hubbard model on bipartite lattice relating total spin to the broken sublattice symmetry [1]. Lieb's theorems on the Hubbard model bipartite lattice is based on two other theorems of Lieb's on the Hubbard model. We will now review all three theorems.

Theorem 6.1 *Lieb's uniqueness theorem for $U < 0$:*
Consider the general Hubbard model:

$$H = \sum_{i,j,\sigma} t_{ij} c^{\dagger}_{i\sigma} c_{j\sigma} + \sum_{i\sigma} U_i n_{i\uparrow} n_{i\downarrow} \tag{6.1}$$

where the elements t_{ij} are assumed to be real. The lattice has no particular symmetry or topology but is connected, i.e., there is a path between any (i, j). Then we have:

If the onsite interactions U_i are all smaller than zero (attractive) and the number of electrons N is even, then $S = 0$ is the unique ground state for any U_i.

Theorem 6.2 *Lieb's half-filled bipartite lattice theorem for $U >> |t|$:*
Now consider a bipartite lattice, i.e., there are A and B types of sites (sublattice) such that $t_{ij} = 0$ if i and j belong to the same sublattice, and with constant U. Then the theorem says:

For a half-filled (number of electrons equals number of sites) bipartite lattice with large and repulsive U, the system is a spin-1/2 Heisenberg antiferromagnet with:

$$H \approx \frac{2}{U} \sum_{i,j} t_{ij}^2 \left(\sigma_i \sigma_j - \frac{1}{4} \right) \tag{6.2}$$

leading to a unique ground state with total spin $2S = N_A - N_B$ where N_A and N_B are number of sites in each sublattice.

Theorem 6.3 *Lieb's general half-filled bipartite lattice theorem:*
The third theorem follows the first two. Consider the constant U Hubbard model for a bipartite lattice. We define the following particle-hole transformation:

$$\tilde{c}_{i\uparrow} = +c_{i\uparrow}^\dagger \quad \text{if} \quad i \in A \tag{6.3}$$

$$\tilde{c}_{i\uparrow} = -c_{i\uparrow}^\dagger \quad \text{if} \quad i \in B \tag{6.4}$$

$$\tilde{c}_{i\downarrow} = +c_{i\downarrow} \quad \text{no change for down spins} \tag{6.5}$$

Then the Hamiltonian in (6.1) becomes

$$\tilde{H} = \sum_{i,j,\sigma} t_{ij} \tilde{c}_{i\sigma}^\dagger \tilde{c}_{j\sigma} - U \sum_{i\sigma} \tilde{n}_{i\uparrow} \tilde{n}_{i\downarrow} + UN_\downarrow \tag{6.6}$$

Now, the Theorem 6.1 implies that the ground state of $\tilde{H}$ is unique for any $U > 0$ (no degeneracy or crossing allowed), thus H also has a unique ground state S. On the other hand, the Theorem 6.2 implies that for $U >> 0$, $2S = N_A - N_B$. We can then conclude that:

For a half-filled bipartite lattice with repulsive U, the ground state is unique and has a total spin $S = |N_A - N_B|/2$ where N_A and N_B are number of sites in each sublattice.

The third theorem has important implications for any graphene nanostructure with zigzag edges, since zigzag edges break the symmetry between the two sublattices resulting in finite magnetism. However, unlike in Lieb's theorem, in our calculations the many-body interacting Hamiltonian contains direct long-range, exchange, and scattering terms. Moreover, we include the next-nearest-neighbor hopping integral in HF self-consistent calculations that slightly violates the bipartite lattice property of the TGQD, one of the cornerstones of Lieb's arguments [1]. Nevertheless, the main result of the spin-polarized ground state for the charge-neutral TGQD seems to be consistent with predictions of Lieb's theorem and, hence, applicable to much larger systems.

Having established the spin polarization of the charge-neutral TGQD we now discuss the spin of charged TGQD. We start with a spin-polarized ground state $|GS\rangle$ of a charge-neutral TGQD with all electron spins down, and add to it a minority spin electron in any of the degenerate shell states i as $|i\rangle = a^{\dagger}_{i,\uparrow}|GS\rangle$. The total spin of these states is $S_z^{max} - 1/2$. We next study stability of such states with one minority spin-up electron to spin-flip excitations by forming three particle states $|lki\rangle = a^{\dagger}_{l,\uparrow}a_{k,\downarrow}a^{\dagger}_{i,\uparrow}|GS\rangle$ with total spin $S_z^{max} - 1/2 - 1$. Here there are two spin-up electrons and one hole with spin-down in the spin-polarized ground state. The interaction between the two electrons and the hole leads to the formation of trion states. We form the Hamiltonian matrix in the space of three-particle configurations and diagonalize it to obtain trion states. If the energy of the lowest trion state with $S_z^{max} - 1/2 - 1$ is lower than the energy of any of the charged TGQD states $|i\rangle$ with $S_z^{max} - 1/2$, the minority spin electron forms a bound state with the spin-flip exciton, a trion, and the spin-polarized state of a charged TGQD is unstable. The trion binding energy, shown in Fig. 6.6, is found to be negative for small systems with $N_{deg} \leq 8$ and positive for all larger systems studied here. The binding of the trion, i.e., the negative binding energy, is consistent with the complete spin depolarization obtained using TB+HF+CI method for TGQD with $N_{deg} \leq 9$ but not observed for $N_{deg} = 11$ (and not observed for $N_{deg} = 10$, not shown here), as shown in Fig. 6.5. For small systems, a minority spin-up electron triggers spin-flip excitations, which leads to the spin depolarization. With increasing size, the effect of the correlations close to the charge neutrality vanishes. At a critical size, around $N_{deg} = 8$, indicated by an arrow in Fig. 6.6, a quantum phase transition occurs, from minimum to maximum total spin.

However, the spin depolarization does not vanish but moves to different filling factors. In Fig. 6.5 we observe that the minimum spin state for the largest structure computed by the TB+HF+CI method with $N_{deg} = 11$ occurs for TGQD charged with additional three electrons. We recall that for TGQD with $N_{deg} = 7$ charged with three additional electrons a formation of a Wigner-like spin polarized molecule was observed, shown in Fig. 6.4. In the following, the differences in the behavior of these two systems, $N_{deg} = 7$ and 11, will be explained based on the analysis of the many-body spectrum of the $N_{deg} = 11$ system.

Figure 6.7 shows the many-body energy spectra for different numbers of electrons for $N_{deg} = 11$ TGQD to be compared with Fig. 6.2 for the $N_{deg} = 7$ structure. Energies are renormalized by the energy gap of a half-filled shell, $N_{el} = 11$ electrons in this case. In contrast to the $N_{deg} = 7$ structure, energy levels corresponding to $N_{el} = N_{deg} + 1$ electrons are sparse, whereas increased low-energy densities of states appear for $N_{el} = N_{deg} + 2$ and $N_{el} = N_{deg} + 3$ electrons. In this structure, electrons are not as strongly confined as for smaller systems. Therefore, for $N_{el} = N_{deg} + 3$ electrons, geometrical effects that lead to the formation of a Wigner-like molecule become less important. Here, correlations dominate, which results in a large low-energy density of states.

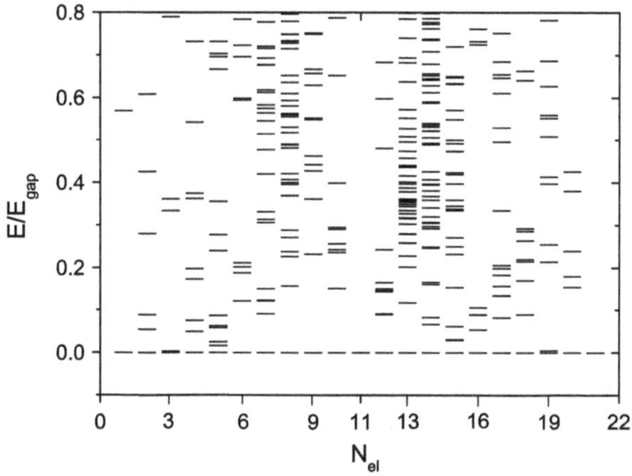

Fig. 6.7 Low-energy spectra of the many-body states as a function of the number of electrons occupying the degenerate shell for the *triangle* with $N_{deg} = 11$ degenerate states. The energies are renormalized by the energy gap corresponding to the *half-filled* shell, $N_{el} = 11$ electrons. The large density of states related to the correlation effects observed in Fig. 6.2 around $N_{deg} + 1$ electrons shifts to a different filling around $N_{edge} + 3$ electrons. Reprinted from [3]

6.1.3 Pair-Correlation Function of Spin Depolarized States

In order to illuminate the depolarization process as an electron is added to the charge-neutral maximally spin polarized system, in Fig. 6.8a, b we show the orbital occupancy of up-spin zero-energy states at $N_{add} - N_{deg} = 1$, for the fully polarized state $S = 3$ (upper panel) and for the ground state, $S = 0$, (lower panel) for the $N = 97$ atoms quantum dot with $N_{deg} = 7$ zero-energy states shown. For the large spin $S = 3$ case, the added spin-up electron simply occupies the orbital 1 and its spin is opposite to the spins of the other 7 electrons. However, the true ground state has $S = 0$, with the spin occupancy shown in the lower panel. The added electron causes electrons already present to partially flip their spin, with spin-up density being delocalized over all the 7 orbitals in analogy to Skyrmion-like excitations in quantum dots and quantum Hall ferromagnets [10–13]. The correlated nature of the $S = 0$ spin depolarized ground state can be investigated using the spin-resolved pair correlation function defined as

$$P_{\sigma\sigma_0}(\mathbf{r}, \mathbf{r_0}) = \frac{1}{N(N-1)} \langle \sum_{i<j} \delta(\mathbf{r} - \mathbf{r}_i)\delta_{\sigma\sigma_i}\delta(\mathbf{r_0} - \mathbf{r}_j)\delta_{\sigma_0\sigma_j} \rangle$$

which is proportional to the probability of finding an electron with spin σ at $\mathbf{r}$ when an electron with spin σ_0 is held fixed at $\mathbf{r}_0$.

In Fig. 6.8c, d, $P_{\uparrow\uparrow}(\mathbf{r}, \mathbf{r_0})$ is shown for $S = 3$ and $S = 0$ states, respectively. The location of the fixed up-spin electron at site $\mathbf{r}_0$ is schematically shown with an

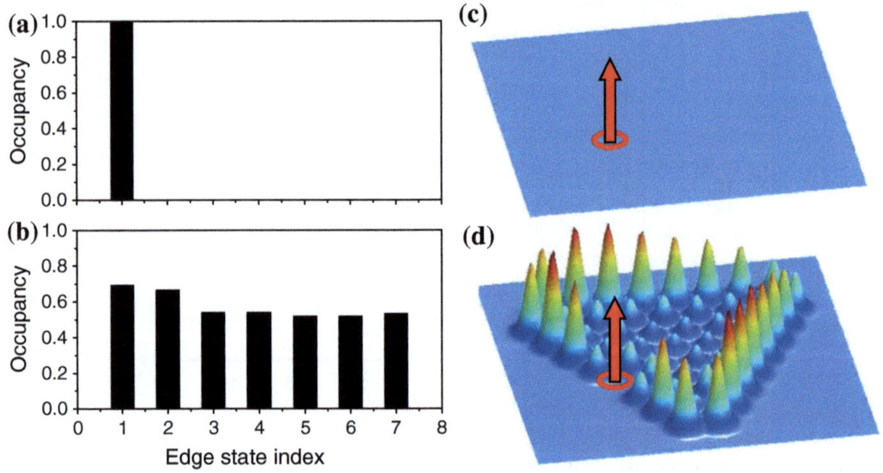

Fig. 6.8 *Left panel* Orbital occupancy of the $N_{deg} = 7$ zero-energy states by spin-up electrons, for the charged ($N_{add} - N_{deg} = 1$) system, for **a** $S = 3$ and **b** $S = 0$ total spin states. The ground state is $S = 0$ (see Fig. 6.2). *Right panel* Corresponding spin up-up pair-correlation functions $P_{\uparrow\uparrow}(\mathbf{r}, \mathbf{r_0})$. The fixed spin-up electron is represented by a *red arrow*, and its position $\mathbf{r}_0$ by a *red circle*. Reprinted from [2]

up arrow. The up-up spin correlation function for the $S = 3$ spin polarized system is strictly zero as there are no spin-up electrons. The spin correlation function for the spin depolarized ground state with $S = 0$ shows the exchange hole at $\mathbf{r}_0$, which extends to the nearest neighbors, and, more interestingly, for larger $|\mathbf{r}_0 - \mathbf{r}|$, the spin pair correlation function reveals a spin texture: Beyond the exchange hole there is the formation of an electronic cloud with positive magnetization which decreases and changes sign at even larger distance, again consistent with the Skyrmion picture [10, 11].

6.1.4 Coulomb and Spin Blockades in TGQD

Experimentally, spin properties of quantum dots can be probed using the Coulomb and spin blockade spectroscopy [14]. By connecting the graphene quantum dot to leads and measuring the conductance as a function of gate voltage, one obtains a series of Coulomb blockade peaks. The relative position of these peaks and their height reveal information about the electronic properties of the system as the number of electrons is increased. The amplitude of the Coulomb blockade peak is given by the conductivity G_i of the graphene quantum dot connected to leads via atom "i" [15] as shown schematically in Fig. 6.9a. Spin and correlation effects are reflected in the weight of the Coulomb blockade peak proportional to the matrix element $|\langle N_{el} + 1, J', S'|c^{\dagger}_{i\sigma}|N_{el}, J, S, \rangle|^2$, which gives the transition probability from state

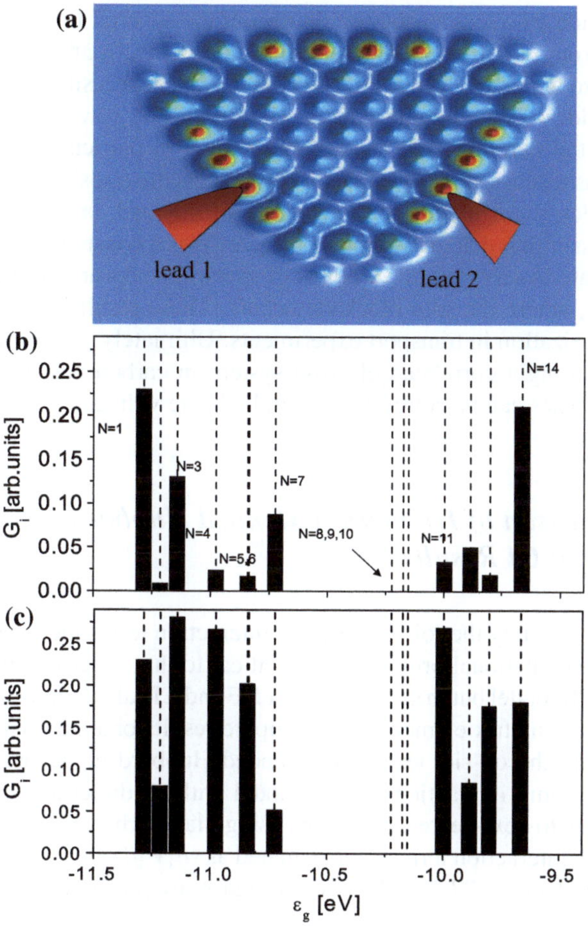

Fig. 6.9 **a** Schematic representation of the graphene island connected to the leads through a side site. **b** Conductivity as a function of the shift in single-particle energies due to applied gate voltage, ϵ_g, (c) Same as (**b**) but without the site dependence of the incoming electron. The oscillations of the spectral weight in (**c**) are purely to due correlation effects and spin blockade. Reprinted from [2]

(N_{el}, J, S) to the state $(N_{el} + 1, J', S')$ when an additional electron is added to the site "i" of the graphene quantum dot from the lead. The ground-state configuration (N_{el}, J, S) is controlled by the gate voltage. For our model graphene quantum dot with $N_{deg} = 7$ degenerate zero-energy states, we can add a total of $N_{add} = 14$ electrons. Hence, one expects to obtain $N_{add} = 14$ Coulomb blockade peaks. In Fig. 6.4b some of the peaks have zero height due to the spin blockade phenomenon. For instance, the transitions from $(N_{el} = 7, S = 7/2)$ states to $(N_{el} = 8, S = 0)$ states are spin blocked since it is not possible to change the spin of the system by $\Delta S = -7/2$ by adding one electron with $S = 1/2$. Similarly, transitions from $(N_{el} = 9, S = 1/2)$ states to $(N_{el} = 10, S = 4/2)$ states are spin blocked. Besides the spin blockade, one sees strong oscillations of the spectral function heights. This is due to (a) strongly-correlated nature of the states $|N_{el}, S\rangle$, and (b) specific choice of the site "i", to which the lead is attached. Here, we chose a site close to the middle of one of the sides of the triangle. The overlap of the site wave function is strongly dependent on the nature of the zero-energy states. In particular, the existence of corner states, as discussed in Fig. 6.1, strongly affects the transition probabilities. To isolate the effect of correlation to the position of the lead, in Fig. 6.9c we plot the conductivity assuming that the weight of the site "i" is a constant independent of the zero-energy state. As a result, the weights of spectral peaks are different, except for $N_{el} = 8, 9, 10$, where the spin blockade occurs. These results show how to detect the spin depolarization in transport experiments. Ultimately, we show here that one can design a strongly correlated electron system in carbon-based material whose magnetic properties can be controlled by applied gate voltage.

6.1.5 Comparison of Hubbard, Extended Hubbard and Full CI Results

In this section we study the role of different interaction terms included in our calculations. The computational procedure is identical to that described in Sect. 3.6. We start from the TB model but in self-consistent HF and CI calculations we include only specific Coulomb matrix elements. We compare results obtained with the Hubbard model with only the on-site term, the extended Hubbard model with on-site plus long-range Coulomb interactions, and a model with all direct and exchange terms calculated for up to next-nearest neighbors using Slater orbitals, and all longer range direct Coulomb interaction terms approximated as $\langle ij|V|ji\rangle = 1/(\kappa|r_i - r_j|)$, written in atomic units, 1 a.u. = 27.211 eV, where r_i and r_j are positions of ith and jth sites, respectively.

The comparison of HF energy levels for the structure with $N_{deg} = 7$ is shown in Fig. 6.10. The on-site U term slightly removes the degeneracy of the perfectly flat shell (Fig. 6.10a) but preserves the double valley degeneracy. On the other hand, the direct long-range Coulomb interaction separates three corner states from the rest with a higher energy (Fig. 6.10b), forcing the lifting of one of the doubly degenerate subshells. Finally, the inclusion of exchange and scattering terms causes a stronger removal of the degeneracy and changes the order of the four lower-lying states. However, the form of the HF orbitals is not affected significantly (not shown here).

In Fig. 6.11 we study the influence of different interacting terms on CI results. The phase diagrams obtained within (a) the Hubbard model and (b) the extended Hubbard model show that all electronic phases are almost always fully spin polarized. The ferromagnetic order for the charge-neutral system is properly predicted. For TGQD charged with electrons, only inclusion of all Coulomb matrix elements correctly predicts the effect of the correlations leading to the complete depolarization for $N_{el} = 8$ and 9. We note that the depolarizations at other filling factors are also

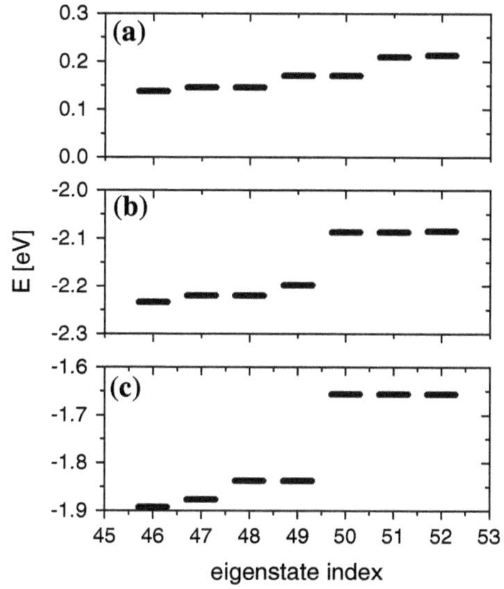

Fig. 6.10 Hartree-Fock energy levels corresponding to the degenerate shell for calculations with **a** only the on-site term U (Hubbard model), **b** the on-site term U + direct long-range interaction (extended Hubbard model), and **c** all interactions. A separation of three corner states with higher energies is related to direct long range Coulomb interaction terms. Reprinted from [3]

Fig. 6.11 Spin phase diagrams obtained by use of the CI method with **a** only the on-site term U (Hubbard model), **b** the on-site term U + direct long range interaction (extended Hubbard model), and **c** all interactions. The ferromagnetic order for the charge-neutral system is properly predicted by all three methods. Correlations leading to the complete depolarization for $N_{el} = N_{deg} + 1$ electrons and $N_{el} = N_{deg} + 2$ electrons are observed only within a full interacting Hamiltonian. Reprinted from [3]

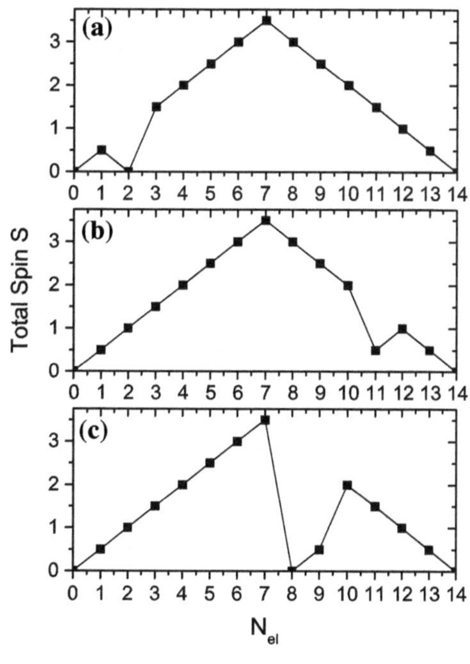

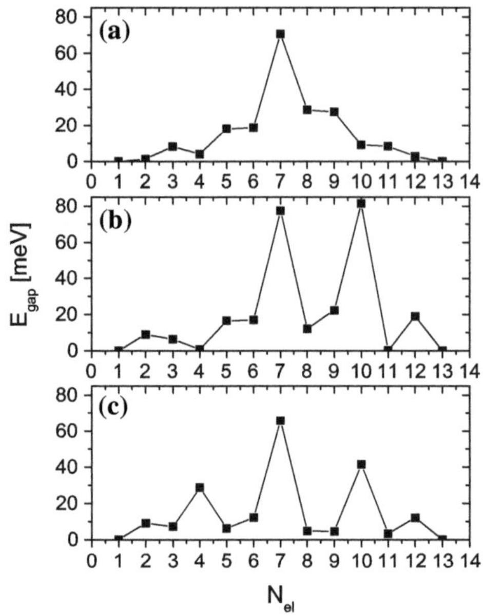

Fig. 6.12 Excitation gaps corresponding to phase diagrams from Fig. 6.11 for many-body Hamiltonians with **a** only the on-site term U (Hubbard model), **b** the on-site term U + direct long-range interaction (extended Hubbard model), and **c** all interactions. All three methods give qualitatively similar excitation gaps for the charge neutral system. A large energy gap for $N_{el} = N_{deg} + 3$ electrons, which is related to geometrical properties of the structure, can be obtained by inclusion of direct long-range interactions. This gap is slightly reduced by inclusion of exchange and scattering terms. Reprinted from [3]

observed in Hubbard (at $N_{el} = 2$) and extended Hubbard calculation (at $N_{el} = 11$) results.

A more detailed analysis can be done by looking at the energy excitation gaps, which are shown in Fig. 6.12. For the charge-neutral system, all three methods give comparable excitation gaps, in agreement with previous results [2, 7–9]. In the Hubbard model, the energy gap of the doped system is reduced compared to the charge neutrality but without affecting magnetic properties. The inclusion of the direct long-range interaction in Fig. 6.12b induces oscillations of the energy gap. For $N_{el} = N_{deg} + 1$ electrons the energy gap is significantly reduced but the effect is not sufficiently strong to depolarize the system. Further away from half-filling, a large energy gap for models with long-range interactions for $N_{el} = N_{deg} + 3$ appears, corresponding to the formation of a Wigner-like molecule of three spin-polarized electrons in three different corners. The inclusion of exchange and scattering terms slightly reduces the gap but without changing the main effect of Wigner-like molecule formation.

6.1.6 Edge Stability from Ab Initio Methods

So far we have assumed that the edges of the quantum dot are perfect, i.e., the edges are not reconstructed from hexagonal shape and there are no dangling bonds. The second assumption can be correct if the system reacts with the right amount of hydrogen which will effectively passivate the dangling bonds. On the other hand, the experiments show that zigzag edges of graphene structure can sometimes lose their hexagonal shape through the so-called pentagon-heptagon reconstruction (Fig. 6.13) [16–20].

We would like to analyze the robustness of the magnetic properties of the TGQD versus edge reconstructions and passivation as a function of size. To do this, we have performed DFT calculations of various TGQD structures using the SIESTA code [21]. We have used the generalized gradient approximation (GGA) with the Perdew-Burke-Ernzerhof (PBE) exchange-correlation functional, [22] double-ζ plus polarization (DZP) orbital basis for all atoms, i.e., 2s, 2p, and 2d orbitals for carbon, thus, both σ and π bonds are included on equal footing, Troullier- Martins norm-conserving pseudopotentials to represent the cores, 300-Ry real-space mesh cutoff for charge density (with symmetrization sampling to further improve the convergence), and a supercell with at least 20 Å of vacuum between the periodic images of the TGQDs. Geometries were optimized until the forces on atoms below 40 meV/Å were reached, and exactly the same geometries were used for the comparison of total energies of the ferromagnetic (FM) and antiferromagnetic (AFM) configurations. Our optimized

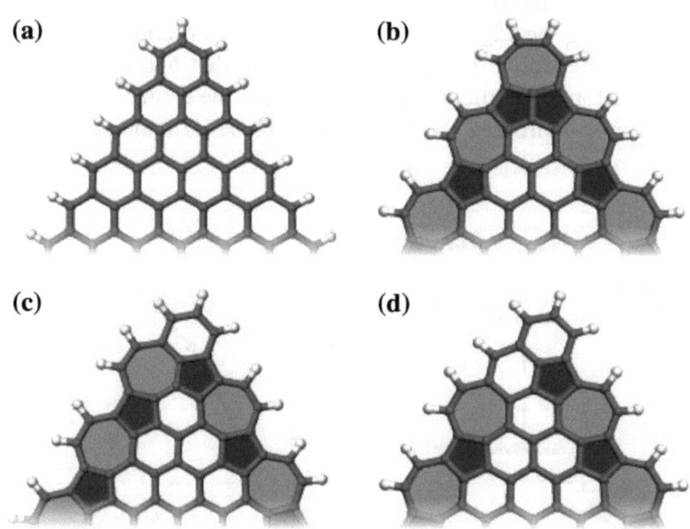

Fig. 6.13 *Triangular* graphene quantum-dot edge configurations considered in this Section: **a** Ideal zigzag edges *ZZ*, **b** ZZ_{57} reconstruction with pentagon-heptagon-pentagon corner configuration, **c** ZZ_{57} reconstruction with heptagon-hexagon-pentagon corner, and **d** ZZ_{57} reconstruction with hexagon-hexagon-pentagon corner. Reprinted from [16]

C-C bond length for bulk graphene of 1.424 Å overestimates the experimental value by ∼3 %, typical for GGA.

We assume the two types of edge structures, zigzag edge (ZZ) and the reconstructed pentagon-heptagon (ZZ_{57}) zigzag edge as shown in Fig. 6.13 for different possible arrangements. The ZZ_{57} reconstruction of the edge leads to structures with the three rings at the corner, 5-7-5, 7-6-5, or 6-6-5 arrangements presented in Fig. 6.13b, d (for the sake of comparison of total energies, we investigate only those reconstructions conserving the number of atoms). Among reconstructed corners, only the structure in Fig. 6.13b conserves the mirror symmetry of the TGQD; however, according to our calculations, it is the least stable due to strong distortion of the corner cells. Thus, in the following we will be presenting results utilizing the configuration shown in Fig. 6.13c for an even, and that in Fig. 6.13d for an odd number of edge atoms on one edge denoted by N_{edge}.

Passivation by hydrogen is an important requirement for the observation of the band of nonbonding states. Our calculations show that, without hydrogen passivation, the π bonds hybridize with the σ bonds on the edge, destroying the condition of the well-defined equivalent bonds on a bipartite lattice and thus the zero-energy band itself. In Fig. 6.14a we address the stability of hydrogen passivation for ZZ and ZZ_{57} edges on the example of a triangle with $N = 97$ carbon atoms, with $N_{edge} = 8$. The number of passivating hydrogens is $N_H = 3N_{edge} + 3 = 27$. For hydrogen-passivated structures, *ZZH* is 0.3 eV per hydrogen atom more stable than $ZZ_{57}H$, since in the latter structure, the angles between the σ bonds significantly deviate from the ideal 120° angle and the total energy is affected by strain. In the absence of hydrogen, however, the structure has to passivate the dangling σ bonds by itself, e.g. by reconstructing the edge. Indeed, the ZZ_{57} reconstruction becomes 0.4 eV per hydrogen atom more stable. It is important to note that hydrogen passivation is a favorable process for both structures, even relative to the formation of H_2 molecules, and not only atomic hydrogen, i.e., formation of the *H–H* bond can not compensate

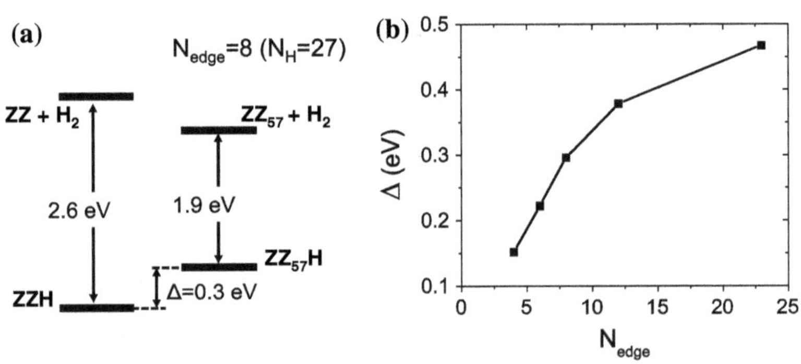

Fig. 6.14 **a** Relative total energies of hydrogen-passivated and nonpassivated TGQDs with reconstructed and non-reconstructed edges for the case of $N_{edge} = 8$ ($N_H = 27$). **b** Total energy difference between hydrogen-passivated ZZ_{57} and ZZ configurations as a function of the number of atoms on a side of the triangle. Presented values are energy per hydrogen atom. Reprinted from [16]

the energy loss due to breaking the *C–H* bond, Fig. 6.14a. The same conclusions hold for larger TGQDs as well. Thus, the H-passivated edge, required for magnetism, is achievable and we will present further only the results for hydrogen-passivated structures omitting the index *H* (i.e., use *ZZ* instead of ZZ_H). These results are also consistent with the ones for infinite edges in graphene nanoribbons [17]. In Fig. 6.14b, we investigate the relative stability of hydrogen-passivated *ZZ* and ZZ_{57} structures as a function of the linear size of the triangles. The largest TGQD that we have studied has $N = 622$ carbon atoms with $N_{edge} = 23$. The fact that the energy per edge atom increases with size implies that the infinite limit has not yet been reached. Clearly, the *ZZ* structure remains the ground state for the range of sizes studied here.

In Sect. 2.2.2 we proved that the number of zero-energy states equals the difference between the number of A and B type of atoms, $N_{deg} = |N_A - N_B|$. Additionally, these states are localized exclusively on the sublattice to which the *ZZ* edges belong, see Fig. 4.19. Figure 6.15 compares the DFT electronic spectra near the Fermi level for the ground states of hydrogen-passivated unreconstructed (*ZZ*) and the reconstructed (ZZ_{57}) TGQDs with $N_{edge} = 12$. The degenerate band survives in a reconstructed (ZZ_{57}) TGQD. However, the dispersion of this band increases almost threefold due

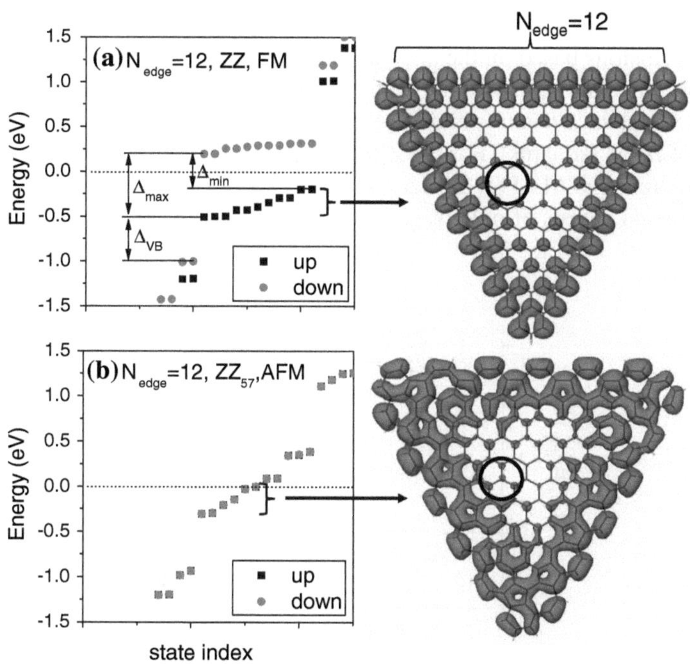

Fig. 6.15 Energy spectra of the ground states for **a** *ZZ* and **b** ZZ_{57} configurations for a hydrogen-passivated triangular dot with $N_{edge} = 12$. Spin-up states are shown in *squares* and spin down states are shown in *circles*. On the right-hand side, charge densities of the filled part of degenerate bands are shown. *Circular outlines* show the population of only one sublattice in the ZZ structure and both sublattices in ZZ_{57}. Reprinted from [16]

to a reduction of the structure symmetry. Lifting of the band degeneracy is observed even in the nearest-neighbor TB model with equal hoppings (not shown), and is more pronounced for the structures in Fig. 6.13c, d, which additionally lift the reflection symmetry present in Fig. 6.13b. For *ZZ* triangle the up- and down-spin edge states are split around the Fermi level such that only up-spin states are filled. This is in agreement with our previous TB+HF+CI results presented in Sect. 4.1. The calculated dispersion of the up-spin states is 0.03 eV/state, see Fig. 6.15a. On the other hand, the ground state of the ZZ_{57} configuration is antiferromagnetic, i.e., there is no splitting between the up- and down-spin states, see Fig. 6.15b. We can understand this in the following way. The introduction of the ZZ_{57} edge reconstruction smears the distinction between sublattices. One can see from the charge-density plot in Fig. 6.15b that degenerate states can now populate both A and B sublattices even close to the center of the dot (see outlined regions), contrary to *ZZ* triangle, see charge-density plot in Fig. 6.15a). We speculate that the resulting reduction in the peak charge density on each site is responsible for the reduced on-site repulsion between spin-up and spin-down electrons. Stronger dispersion and reduced up-down spin splitting favor kinetic energy minimization versus exchange energy and destroy the ferromagnetism in ZZ_{57}. It should be noted that partial polarization can still be possible in ZZ_{57}. Particularly, we observed it for structures with symmetric corners (Fig. 6.13b), which exhibit smaller dispersion.

Our conclusions based on the analysis of the energy spectra are supported by the total energy calculations shown in Fig. 6.16. For the *ZZ* structure, the gap Δ_{min} shown in Fig. 6.15 is always positive and the total energy of the FM configuration is lower than that of AFM (squares). For the ZZ_{57} configuration, on the contrary, the ground state clearly remains AFM for all sizes with the exception of the case with $N_{edge} = 4$. Here, the band consists of only $N_{deg} = 3$ degenerate states and their dispersion can

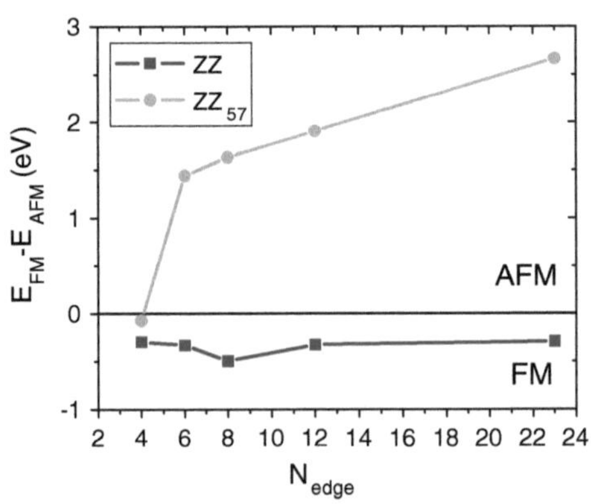

Fig. 6.16 Total energy difference between ferromagnetic and antiferromagnetic states as a function of the size of the *triangle* for hydrogen-passivated *ZZ* (*squares*) and ZZ_{57} (*circles*). For *ZZ*, the ground state is ferromagnetic for all sizes studied, while for ZZ_{57} it is antiferromagnetic for $N_{edge} >$ 4. Reprinted from [16]

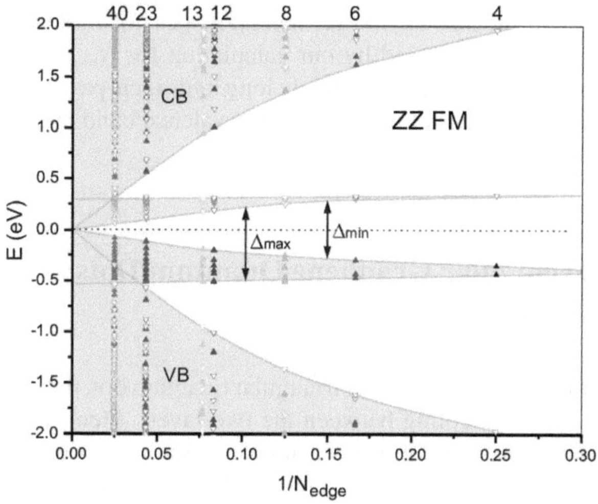

Fig. 6.17 Scaling of the energy gaps with the inverse linear size of *ZZ* TGQDs. Full energy spectra of the structures calculated in this work are shown. *Open and filled symbols* correspond the full energy spectra for spin-down and spin-up states, respectively. Reprinted from [16]

not overcome the splitting between spin-up and spin-down states, resulting in the FM configuration being more stable. The total energy difference between the FM and AFM configurations for *ZZ* remains almost constant (in the range 0.3–0.5 eV) for the triangle sizes studied here, and reduces with size if divided by the number of edge atoms. Such a small value, comparable to the numerical accuracy of the method, makes it difficult to make reliable predictions regarding magnetization of larger dots.

To investigate whether the magnetization of the edges would be preserved on a mesoscale, we plot in Fig. 6.17 the evolution of the energy spectra with the TGQD size. For this plot, we performed an additional calculation for the case of N = 1,761 carbon atoms with $N_{edge} = 40$. We did not perform the geometry optimization for this case due to the high computational cost, however, based on the results for smaller structures, we expect that this would have a minor effect on the spectrum. This allows us to observe the reduction of the splitting Δ_{max} shown in Fig. 6.15 between the spin-up and spin-down states with the growing size, which was not appreciated previously. Our GGA gap between degenerate bands (Δ_{min}) and that between the valence and conduction bands are larger than LDA gaps reported previously, as also observed for graphene nanoribbons. Both gaps show sublinear behavior, complicating the extrapolation to triangles of infinite size. This behavior, however, should change to linear for larger structures where the effect of edges reduces, converging both gaps to zero, as expected for Dirac Fermions. An important difference from the nearest-neighbor TB calculation is the growing dispersion of the zero-energy bands. Combined with the reduction of the valence-conduction gap, this leads to the overlap of the degenerate band with the valence band, even for finite sizes, as indeed observed for the $N_{edge} = 40$ case (see Fig. 6.17), while in ZZ_{57} structures, it becomes visible already at $N_{edge} = 23$ (not shown). Nevertheless, it does not affect the magnetization of the edges, as indeed confirmed by our calculation for $N_{edge} = 40$, and can be compared to a magnetization of the infinitely long hydrogen-passivated nanoribbons, where the edge state overlaps in energy with the valence band but, in k space, those bands do not actually cross.

6.2 Bilayer Triangular Graphene Quantum Dots with Zigzag Edges

In Sect. 4.4 we showed that in a bilayer triangular quantum dot, the zero-energy states are not affected by the coupling between the two layers. Hence, we have two sets of zero-energy states originating from each layer. Moreover, we have seen that it is possible to control the relative energies of two sets of zero-energy states by applying an external electric field. In this section we will discuss the magnetic properties of edges and show that the ability of controlling the relative position of the energy of the bilayer graphene quantum dot gives an interesting opportunity to control the charge and spin of the zero-energy states. Most calculations in this section were performed using the mean-field extended Hubbard approximation. In all calculations the on-site Hubbard term U is taken to be 2.75 eV, screened by a factor of ∼6 from the bare Coulomb potential.

In Fig. 6.18 we show the spin density isosurfaces for zero electric field (left hand side) and finite electric field (right hand side), as obtained from configuration interaction calculations. When the electric field is off, both layers have a finite magnetic moment, as in single layer triangles [2, 8, 9, 16, 24–26], differing by one spin due to the size difference of the two triangles. The magnetic moments of the two layers are coupled ferromagnetically, in agreement with Lieb's theorem [1] which applies for Bernal stacking (the edge atoms of the two triangles belong to the same sublat-

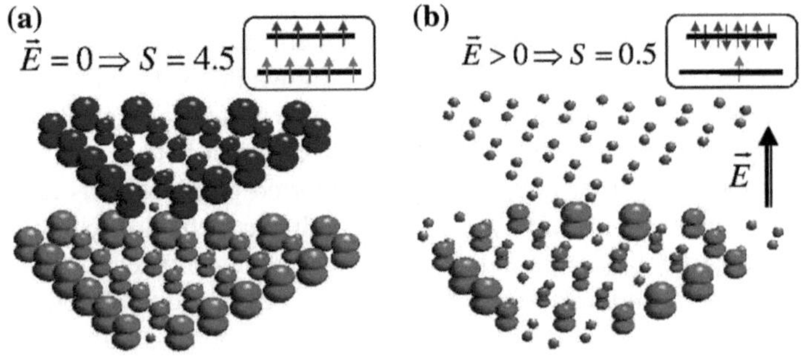

Fig. 6.18 Isosurface plot of the spin density $\rho_\uparrow - \rho_\downarrow$ of a bilayer *triangular* graphene quantum dot with zigzag edges, **a** in the absence and **b** in the presence of a perpendicular electric field obtained from configuration interaction calculations. Reprinted from [23]

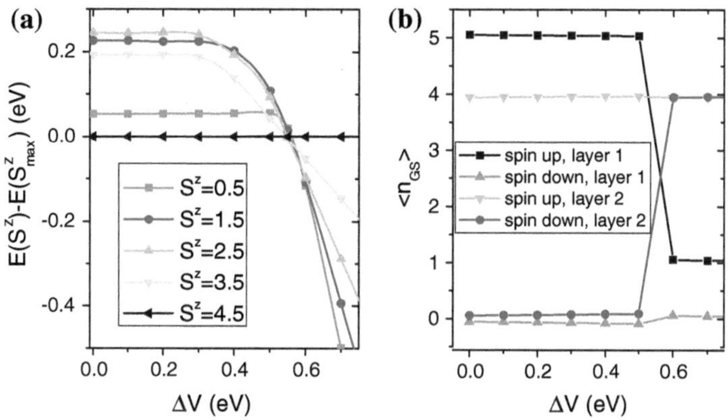

Fig. 6.19 Mean-field Hubbard results for a bilayer graphene quantum dot with 107 atoms and 9 zero-energy states. **a** Energies of lowest energy states with different total spin projection S_z as a function of potential difference ΔV_c between the layers, with respect to the ferromagnetic configuration $S_z^{max} = 4.5$. **b** Ground state spin population for a given layer and spin. Reprinted from [23]

tice). When a sufficiently high electric field is applied, electrons from the lower layer reduce their energy by transferring to the top layer, filling up all the available spin-up and down zero-energy states, leaving behind one single spin. It should be noted that the depolarization effect described above is robust against defects, since Lieb's theorem [1] guarantees magnetization of individual layers and a ferromagnetic coupling between them, as long as the biparticity of the honeycomb lattice is not distorted.

In Fig. 6.19, we study a bilayer triangular quantum dot with 107 atoms and 9 zero-energy states. Figure 6.19a shows the energies for different total spin projection S_z with respect to the energy of the ferromagnetic configuration, $S_z = 9/2$. At $\Delta V = 0$, the degenerate band of zero-energy states is polarized: all 9 electrons occupying the 9 zero-energy states have their spins aligned ferromagnetically as explicitly shown in Fig. 6.18a. Although the first excited state obtained from the Hubbard model is antiferromagnetic with the polarization of the bottom layer opposite to the polarization of the top layer, a full treatment of the correlation effects shows that low lying excited states have more complex spin structures [23]. The Hubbard model is, however, useful for estimating the critical value V_c where the phase transition occurs. As ΔV is increased, the electrons lying on the bottom layer zero-energy states are forced to flip their spin and tunnel to the top layer zero-energy states. At around $\Delta V_c = 0.55$ eV such charge transfers occur abruptly, leading to a decrease of the magnetization. As a result, all top layer zero-energy states become doubly occupied, leaving exactly one single spin in the bottom layer zero-energy states. We note that one can also isolate a single hole spin in the bottom layer by applying a reverse electric field, thus pushing the electrons from the top layer to the bottom layer, occupying all states except one. It is thus possible to isolate a single electron or hole spin in a neutral bilayer graphene quantum dot isolated from metallic leads by applying an external electric field.

The procedure of isolating a single electron or hole should occur regardless of the size of the system, since the top layer has always one fewer zero-energy state

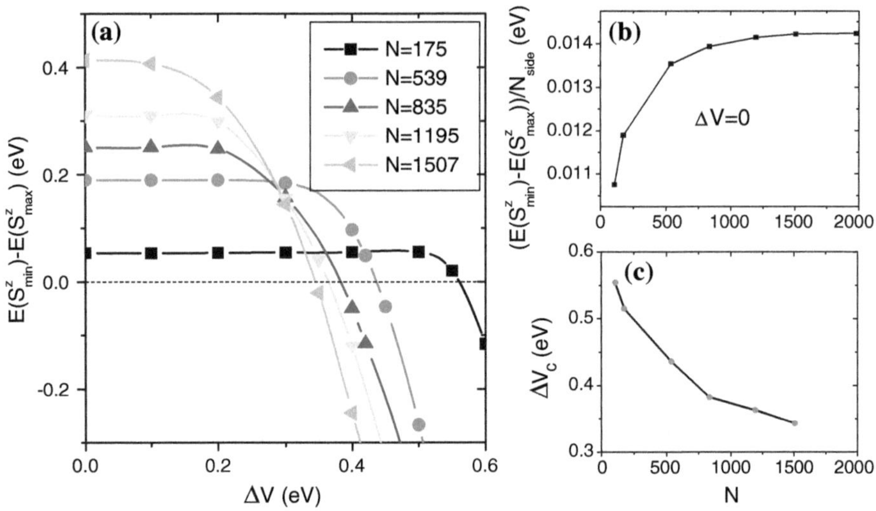

Fig. 6.20 Size dependence of ferromagnetic-antiferromagnetic transition. **a** FM-AFM energy difference as a function of potential difference ΔV, up to $N = 1,507$ atoms. **b** For $\Delta V = 0$, the FM-AFM energy gap per number of side atoms N_{side} approaches 14.3 meV. **c** Critical value ΔV_c where the transition occurs as a function of number of atoms N. Reprinted from [23]

than the bottom one. In order to investigate the size dependence, in Fig. 6.20a we show the energy difference between the ferromagnetic and antiferromagnetic (FM-AFM) states calculated in the mean-field Hubbard approximation as a function of applied voltage for several sizes up to 1,507 atoms. It should be noted that, due to the unusually high degeneracy of the states, self-consistent iterations occasionally get trapped in local energy minima. Hence, it is important to repeat the calculations several times using different initial conditions and/or convergence schemes to assure that the correct ground state was reached. As expected, at $\Delta V = 0$, the FM-AFM gap increases with the size of the system N. In fact, the FM-AFM gap energy per N_{side}, the number of side atoms on the top layer (N_{side}) (equal to the number of inter-layer bonds on the edges), approaches a constant value of 14.3 meV as shown in Fig. 6.20b. However, the FM-AFM transition voltage V_c decreases with the system size as can be seen from Fig. 6.20c. For the largest system size studied, $N = 1{,}507$, we obtain $\Delta V_c = 0.345$ eV, which corresponds to an electrical field of ~1 V/nm, a value within experimental range [27].

6.3 Triangular Mesoscopic Quantum Rings with Zigzag Edges

In Sect. 4.5 we have seen that Triangular Graphene Quantum Rings (TGQRs) have the interesting property that while their outer edges are built of A-type of atoms, their inner edges are built of B-type of atoms, all contributing to zero-energy states. Here we will see that there is again a finite total magnetization in the structure, but the inner and outer edges have opposite polarization, with an antiferrimagnetic coupling between them [28].

6.3.1 Properties of the Charge-Neutral TGQR

In order to study magnetic properties of TGQRs, we use the Hubbard model with the Hamiltonian given by (5.23) which allows to investigate mesoscopic size structures. In order to check the validity of the Hubbard model, we first compare the results with DFT calculations for a smaller sized TGQR containing 171 atoms. Figure 6.21 shows spectra obtained (a) from the Hubbard model in the mean-field approximation and (b) using DFT implemented in SIESTA package [21] for TGQR with $N_{width} = 2$, consisting of $N = 171$ atoms, $N_{out} = 11$ and $N_{inn} = 2$. This corresponds to $N_{deg} = 9$ degenerate zero-energy TB levels, shown in Fig. 4.31a. Interactions open a spin-dependent gap in the single-particle zero-energy shell, resulting in maximum spin polarization of those states. The total spin is $S_{tot} = 9/2$, in accordance with Lieb's theorem [1]. In Fig. 6.21c, d we show the corresponding spin density. The net total spin is mostly localized on the outer edge and vanishes as one moves to the center, similar to the

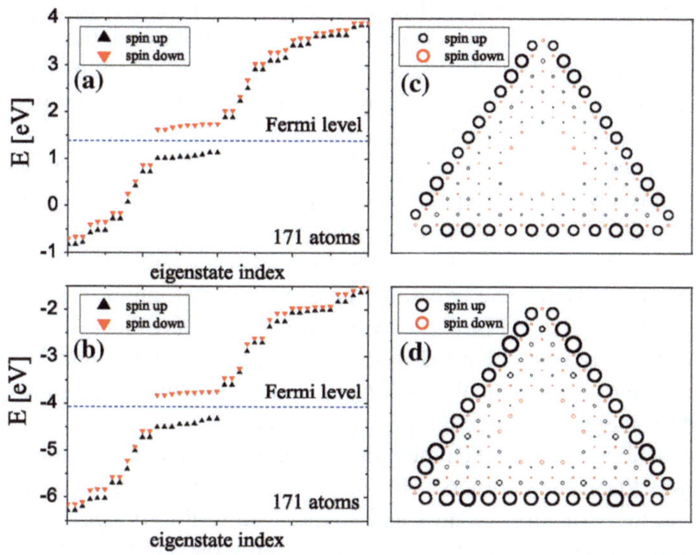

Fig. 6.21 Energy spectra from **a** self-consistent mean-field Hubbard model and **b** DFT calculations for TGQR with the width $N_{width} = 2$ and $N = 171$ atoms. States up to the Fermi level (*dashed line*) are occupied. **c** and **d** are corresponding spin densities. The *radius* of *circles* is proportional to the value of spin density on a given atom. Proportions between size of *circles* in (c) and (d) are not retained. Reprinted from [28]

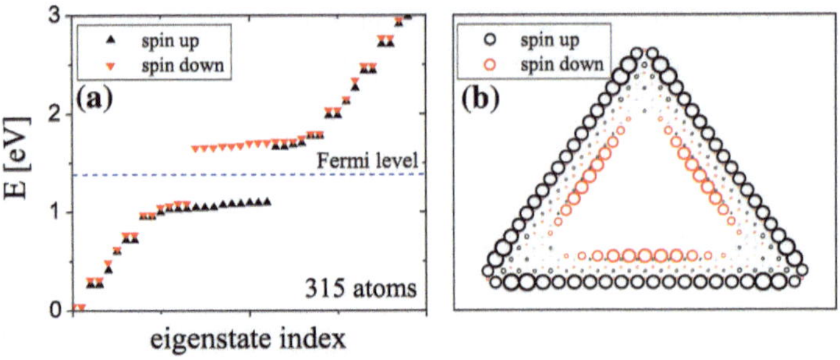

Fig. 6.22 **a** Self-consistent energy spectra and **b** corresponding spin densities from mean-field Hubbard model for TGQR with the width $N_{width} = 2$ and $N = 315$ atoms. The radius of *circles* is proportional to the value of spin density on a given atom. Reprinted from [28]

electronic densities of TGQD shown in Fig. 4.19f. Good agreement between results obtained from the mean-field Hubbard and DFT calculations (Fig. 6.21) validates the applicability of the mean-field Hubbard model and allowed us to study efficiently the structures consisting of a larger number of atoms.

In Fig. 6.22 we show the results of the Hubbard model for a larger structure with $N = 315$ atoms, $N_{out} = 20$ and $N_{inn} = 11$, with the same width $N_{width} = 2$. The energy spectrum, Fig. 6.22a, looks similar to that from Fig. 6.21a and the total spin is again $S_{tot} = 9/2$. On the other hand, the spin density in Fig. 6.22b is different than in Fig. 6.21c. Here, the outer edge is still spin polarized, but the inner edge reveals opposite polarization. This fact can be understood in the following way. Electrons with majority spin (spin up) occupy degenerate levels of the zero-energy shell which are built exclusively of orbitals localized on atoms belonging to the sublattice labeled as A. These states are localized on the outer edge. Due to the repulsive on-site interaction, spin-up electrons repel minority spin electrons (spin down) to the sublattice labeled as B. After self-consistent calculations, spin-up and spin-down densities are spatially separated occupying mostly sublattice A and sublattice B, respectively. Local imbalance between the two sublattices occurs near edges, resulting in local magnetic moments, seen in Fig. 6.22b. As a result, we observe that the outer and inner edges are oppositely spin polarized, similar to graphene nanoribbons. However, the magnetic moments are not equal, resulting in local antiferrimagnetic state in contrast to the antiferromagnetic state in graphene nanoribbons.

The magnetic moment of the inner edge is highest close to the middle of the edge and decreases toward the corners. This allows us to distinguish between two types of regions in the structure: corners and edges. Due to the triangular symmetry of the system, in further analysis we can focus on only one corner and one edge. We define the average magnetization in a given region as $<M>= \sum_i^{'} M_i/N'$, where the summation is over sites in a given region and N' is the corresponding total number of atoms. In Fig. 6.23a we show the average magnetization in one corner and one edge as a function of the size of TGQR for a given width, $N_{width} = 2$. Small structures ($N < 200$ atoms) reveal finite and comparable magnetic moments in both regions,

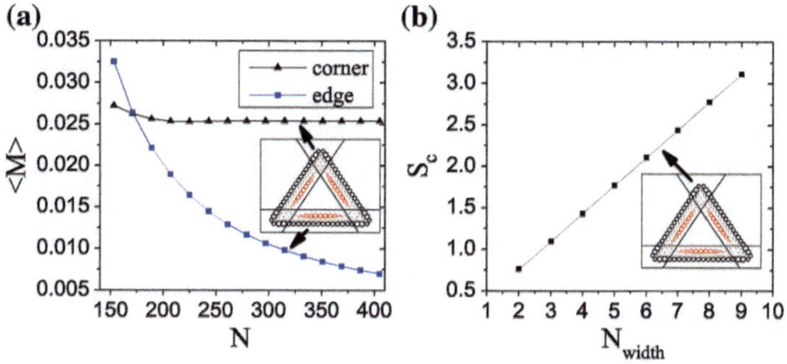

Fig. 6.23 **a** Average magnetic moment as a function of size (N—number of atoms) in corner and edge regions. Structures reveal stable ferromagnetic order in corners, but a change from ferromagnetic to antiferromagnetic on edges with increasing size. **b** Total spin in corner region as a function of width. Linear dependence is due to increased number of zero-energy states. Reprinted from [28]

consistent with Fig. 6.21c, where most of the spin density is distributed on outer edges. There are two effects related to increasing size: the length of the internal edge increases increasing the spin polarization opposite to the outer edge spin polarization (see Fig. 6.22b) and increase of the overall number of atoms in the edge region. The first effect leads to the antiferrimagnetic coupling between opposite edges and the second one to vanishing the average magnetization, seen in Fig. 6.23a. We note here that although the average magnetization rapidly decreases with size, it never approaches zero. On the other hand, average magnetization at the corner is stable and nearly independent of the size. This fact is related to the fixed number of atoms in the corner region.

According to Lieb's theorem [1], the total spin of the system must be $S = 3(N_{width} + 1)/2$. Moreover, the spin density for smaller structures is equally distributed along the outer edge (see Fig. 6.21c). Partitioning the structure into six approximately equal regions, three corners and three edges (see inset in Fig. 6.23a), gives approximately equal total spin in each domain. In further analysis we show that this is true for arbitrary size triangular rings. In Fig. 6.23b we present the total spin in one corner $S_c = \sum_i' M_i$ as a function of the width of the ring. The summation is over all sites in one corner. We obtain a linear dependence $S_c \sim N_{width}$, which for the best choice of cuts should be described by the relation $S_c = (N_{width} + 1)/4$, which is one–sixth of the total spin S of the entire structure. In this ideal case, all six regions reveal equal total spin S_c, independently of the size of the structure. We relate this fact to the behavior in the edge and corner regions. For sufficiently large structures, magnetic moments in the edge region are distributed on a large number of atoms, giving a vanishing average magnetic moment but always a finite total spin equal to $S_c = (N_{width} + 1)/4$. With increasing size, the length of the inner edge increases. In order to satisfy the relation $S_c = (N_{width} + 1)/4$, the magnetic moment on the outer edge increases proportionally to the oppositely polarized magnetic moment on the inner edge, resulting in an antiferromagnetic coupling between opposite edges. On the other hand, in corners, there is always a fixed number of atoms independent of size, giving a constant average magnetic moment and the total spin equal to $S_c = (N_{width} + 1)/4$. We note that above conclusions were confirmed by investigation of TGQR with width in the range $2 \leq N_{width} \leq 9$ for structures up to $N = 1{,}500$ atoms. Thus, we can treat large TGQR as consisting of three ferromagnetic corners connected by antiferrimagnetic ribbons, with ribbons exhibiting a finite total spin. This result can be useful in designing spintronic devices. Choosing TQGRs with proper width, one can obtain a system with a desired magnetic moment localized in the corners.

6.3.2 Filling Factor Dependence of Mesoscopic TGQRs

In the previous section we have shown that the Hubbard model and DFT calculations describe well the properties of the charge-neutral system. On the other hand, in Sect. 6.1 it was shown that gated TGQDs reveal effects related to electronic correlations in the partially filled zero-energy shell [2]. We expect a similar behavior in TGQRs. Thus, in this Section we use again the TB+HF+CI method described in Sect. 3.6 to analyze the magnetic properties as a function of the number of electrons filling the degenerate shell.

We concentrate on the structure shown in Fig. 4.30, consisting of $N = 171$ atoms and characterized by $N_{width} = 2$, which correspond to $N_{deg} = 9$ degenerate states. In Fig. 6.24a we show an example of a configuration related to $N_{el} = 10$ electrons. This corresponds to a half-filled degenerate shell with all spin-down states of the shell filled and an additional spin-up electron. For the maximal total spin $S = 4$ there are nine possible configurations corresponding to the nine possible states of the spin-up electron. The energy spectrum obtained by diagonalizing the full many-body Hamiltonian, (5.29), for total spin $S = 4$ is shown in Fig. 6.24c. We see that, by the comparison with total spin states with $S = 0, 1, \ldots, 4$, the ground state corresponding to configurations of the type a (one of which is shown in Fig. 6.24a) is maximally spin polarized, with the excitation gap in the $S = 4$ subspace of $\sim$40 meV. However, the lowest energy excitations correspond to spin flip configurations with total spin $S = 3$, one of which is shown Fig. 6.24b. These configurations involve spin-flip excitations from the fully spin polarized electronic shell in the presence of the additional spin-up electron.

The energy $E_{gap} = 4$ meV for $N_{el} = 10$, indicated by the arrow in Fig. 6.24c, is shown in Fig. 6.25a together with the energy gap for all electron numbers $1 < N_{el} < 18$ and hence all filling factors. In Fig. 6.25b we show the total spin S of the ground and the first excited state as a function of the number of electrons occupying the degenerate shell. For arbitrary filling, except for $N_{el} = 2$, the ground state is maximally spin polarized. Moreover, the first excited state has total spin consistent with spin-flip excitation from the maximally spin polarized ground state as discussed in detail for $N_{el} = 10$. The signature of electronic correlations is seen in the dependence of the excitation gap on the shell filling, shown in Fig. 6.25a. For the

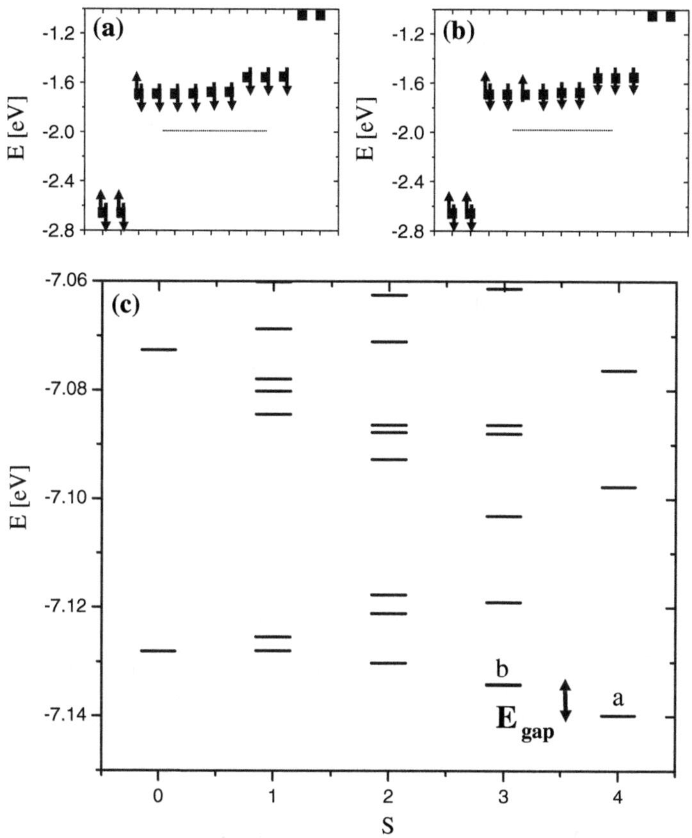

Fig. 6.24 (a) and (b) Hartree-Fock energy levels for TGQR with $N_{width} = 2$ consisting of $N = 171$ atoms and filled by $N_{el} = 10$ electrons. The configuration represented by *arrows* in **a** corresponds to all occupied spin-down orbitals and one occupied spin-up orbital. The configuration represented by *arrows* in **b** is the configuration from (a) with one spin down flipped. **c** The low-energy spectra for the different total spin S for $N_{el} = 10$ electrons. The ground state has $S = 4$, indicated by a, with one of the configuration shown in (a). The lowest energy excited state, indicated by b, is ~4 meV higher in energy, corresponds to spin-flip configurations with one of the configuration shown in (b). Reprinted from [28]

half-filling at $N_{el} = 9$, indicated by an arrow, the excitations are spin-flip excitations from the spin polarized zero-energy shell. This energy gap, ~28 meV, is significantly larger in comparison with the energy gap of ~4 meV for spin flips in the presence of additional spin up electron. The correlations induced by additional spin up electron lead to a much smaller spin-flip excitation energy. This is to be compared with TGQDs where spin-flip excitations have lower energy leading to a full depolarization of the ground state, what was shown in Sect. 4.1 [2].

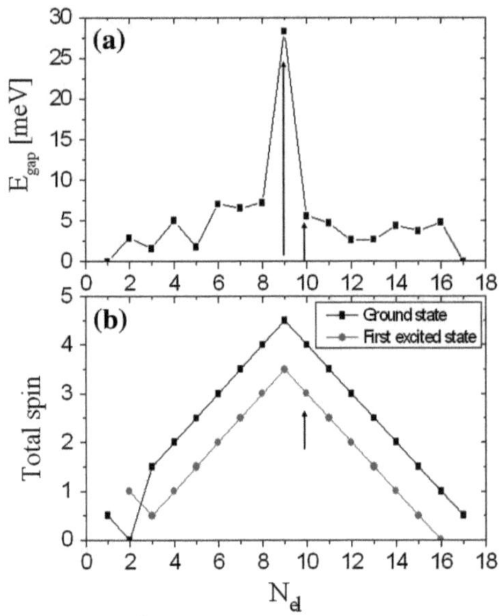

Fig. 6.25 **a** Energy spin gap between ground and first excited state. *Long arrow* corresponds to half-filled shell with $E_{gap} \sim 28$ meV. Significant reduction in the spin flip energy gap for one additional electron, $E_{gap} \sim 4$ meV, indicated by the small *arrow*, is the signature of electronic correlations. **b** Total spin of the ground and first excited state as a function of the number of electrons N_{el}. The small *arrow* indicates excited state for $N_{el} = 10$ electrons with one of the configurations shown schematically with *arrows* in Fig. 6.24b. Reprinted from [28]

6.4 Hexagonal Mesoscopic Quantum Rings

In this section we analyze the size and filling factor dependence of hexagonal quantum rings described in Sect. 4.6. We will see that while the total spin of the ring is minimized for thin rings, there is a critical ring width above which a stable finite magnetization appears. Analysis of the gap as a function of number electron in the structure reveals strong electronic correlations [29].

6.4.1 Dependence of Magnetic Moment in Hexagonal GQRs on Size

In this section we study the ground and excited states as a function of the number of additional interacting electrons in degenerate shells of hexagonal quantum rings with different size L and $W = 1$. Figure 6.26 shows the low-energy spectra for the different total spin S of the half-filled first shell above the Fermi energy for two thinnest rings with a) $L = 4$ and $N = 96$ atoms and b) $L = 8$ and $N = 192$ atoms. For the smaller ring the ground state has total spin $S = 1$ with a very small

Fig. 6.26 Low energy spectra for the different total spin S of the half-filled first shell over the Fermi energy for two thinnest rings $W = 1$ with **a** $L = 4$ and $N = 96$ atoms and **b** $L = 8$ and $N = 192$ atoms. Reprinted from [29]

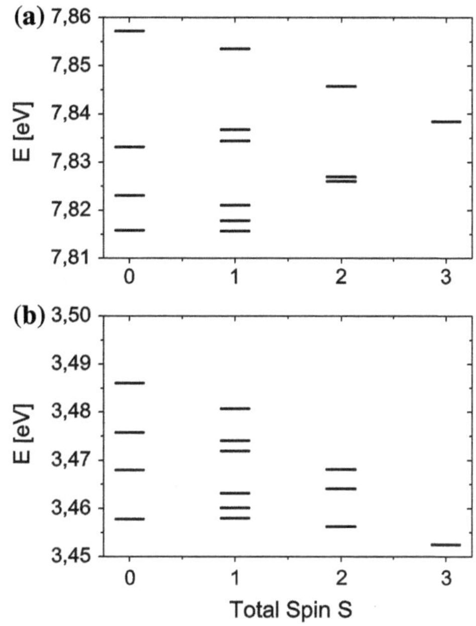

gap to the first excited state with $S = 0$. The lowest states with larger total spin have higher energies. For the ring with $N = 192$ atoms the total spin of the ground state is maximal, $S = 3$. The lowest levels with different total spin have slightly higher energies. This can be understood in the following way. The energy splitting between levels is large for smaller structures, which is seen in Fig. 4.36. For the ring with $L = 4$ and $N = 96$ atoms this value, 0.17 eV, is comparable with electronic interaction terms, e.g., 0.34 eV for two electrons occupying the lowest state. For the ring with $L = 8$ and $N = 192$ atoms the electron-electron interaction terms are 0.23 eV for interaction between two particles on the first state, which is much larger than the single-particle energy difference 0.015 eV. From this, we clearly see that for the ring with $L = 4$ it is energetically favorable to occupy low energy states by electrons with opposite spins. For the ring with $L = 8$ all states have similar energies and due to exchange interactions the lowest energy state is maximally spin polarized.

The behavior of total spin of the ground state for the half-filled shell as a function of size is shown in Fig. 6.27. In this case, the ground state spin can be explained as a result of the competition between occupation of levels with smallest single-particle energies which favors opposite spin configurations, and parallel spin configurations for which exchange interactions are maximized. For rings with $L \geq 5$ the ground state is maximally spin polarized. Here, the splitting between levels is relatively small and the ground state is determined by electronic interactions. Moreover, this splitting decreases with increasing size and this is seen in the spin gap (defined here as the energy required to change the spin of the ground state) behavior (Fig. 6.27). The largest spin gap is observed for ring with $L = 6$ and it decreases with increasing L. For

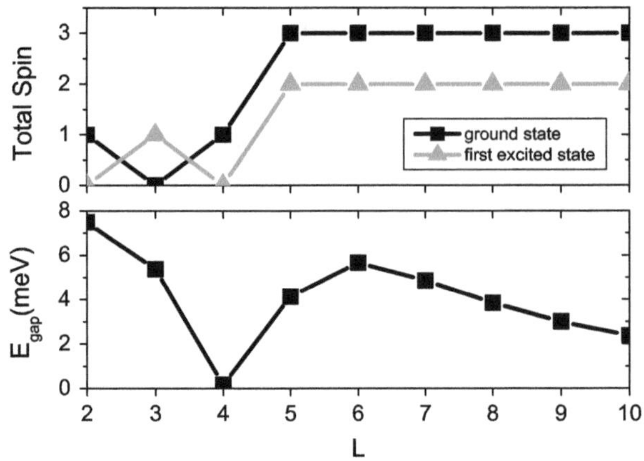

Fig. 6.27 *Upper* Total spin of the ground and first excited states for the half-filling of the first shell in the thinnest ring structures $W = 1$ with different sizes. *Lower* Corresponding energy gap between ground and first excited states with different spin. Reprinted from [29]

small rings the situation is more complicated. Here, the contributions from single particle energies and interactions are comparable. As a consequence, we observe ground states with alternating total spin $S = 1$ and $S = 0$. For sufficiently large rings, $L > 5$, we observe a stabilization of the spin phase diagram. This is connected to changes of the energy differences between levels in a shell—above a critical size these values are so small that they do not play a role anymore.

6.4.2 Analysis as a Function of Filling Factor

In Fig. 6.28 we show the phase diagram for a ring with $L = 8$ and $N = 192$ atoms. Near the half-filling the ground state is maximally spin polarized, which is related to the dominant contribution from the short-ranged exchange interaction terms, and the charge density is symmetrically distributed in the entire ring (see Fig. 4.35). Adding or removing electrons causes irregularities in the density distribution, and correlation effects start becoming important. This results in an alternating spin between maximal polarization (e.g. 3, 4, 9 extra electrons) and complete depolarization (e.g. 2, 8, 10 extra electrons) of the system.

6.5 Nanoribbon Rings

Graphene nanoribbons (see Sect. 4.2.3) played an important role in inspiring the field of topological insulators [30, 31]. The interior of the graphene ribbon acts as an insulator with a gap in the energy spectrum, whereas the energies of the edge states

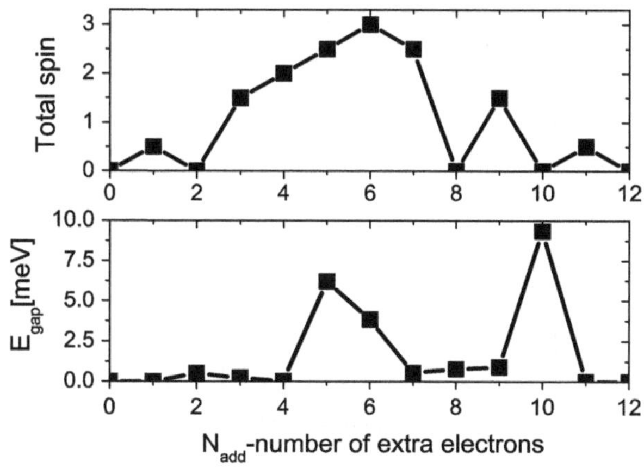

Fig. 6.28 *Upper* The spin phase diagram for electrons occupying the first shell over the Fermi level of the ring structure with $L = 8$ and $N = 192$ atoms. *Lower* Corresponding energy spin gap between ground and first excited states

are in the middle of the gap. The topological aspect of the edge states is generated by the spin-orbit coupling which lifts the spin degeneracy at a given edge, leading to graphene nanoribbon acting as a spin Hall insulator [30]. Unfortunately, the spin-orbit coupling in graphene is found to be too small to give rise to the spin Hall effect [32]. Interestingly, it has been suggested that nontrivial properties of graphene nanoribbons can be generated directly by engineering a nontrivial Möbius geometry of the nanoribbon without the need for the spin-orbit coupling [33–41].

In this section we investigate magnetic properties of graphene nanoribbon rings and compare the Möbius topological insulator with normal insulators, the cylindrical graphene nanoribbon rings without a twist, i.e., with cyclic boundary conditions. We also show that the magnetic properties of the Möbius ribbon have similarities with triangular graphene quantum dots with zigzag edges [2, 3]. Indeed, both systems have only a single edge and insulating bulk, and we compare the magnetic properties as a function of the filling of the edge states with carriers in both structures.

Figure 6.29 shows the result obtained for a cyclic ribbon with length $M = 26$ for different widths $N = 2$, 8, and 14. The system has eight edge states which can be occupied by up to $N_e = 16$ electrons. When the system is charge neutral, i.e., $N_e = 8$, we find that the cyclic ribbon has minimal total spin $S = 0$. For the wide ribbon, $N = 14$, this result is consistent with infinite ribbon results where opposite edges are in an antiferromagnetic configuration carrying opposite spin, with a zero net magnetization. However, here we find that this result is sensitive to the net charge in the system. If we charge the ribbon with even a single electron or hole, an abrupt change from antiferromagnetic configuration to ferromagnetic configuration occurs. In fact, away from the charge neutrality, the total spin of the edges is maximized. When $N_e = 0$ or 16 we have completely empty or doubly occupied edge states, leading to total zero energy again in a paramagnetic configuration. In the other

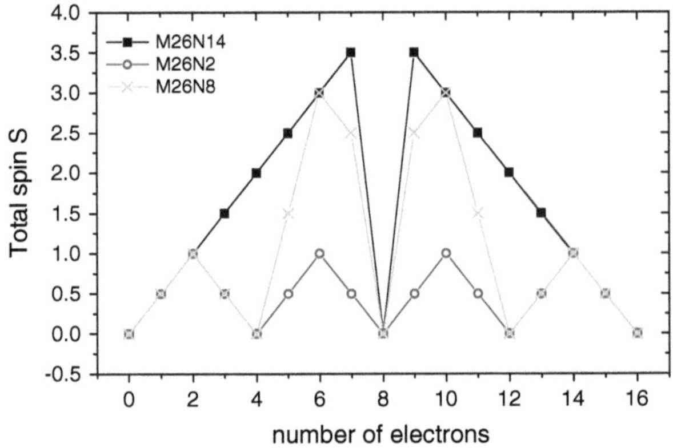

Fig. 6.29 Total spin S as a function of number of electrons occupying the edge states for a cyclic ribbon with length $M = 26$ for different widths, $N = 2$, 8, and 14. Reprinted from [33]

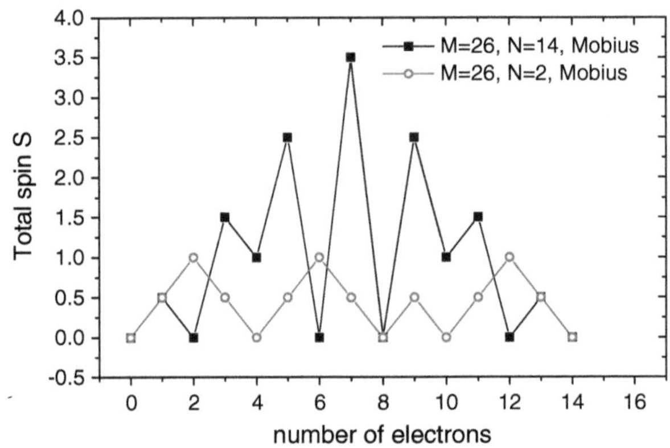

Fig. 6.30 Total spin S as a function of number of electrons occupying the edge states for a Möbius ribbon with length $M = 26$ for different widths, $N = 2$ and 14. Reprinted from [33]

limit of thin ribbon, $N = 2$, the edge states are not highly degenerate but form a shell structure with double orbital degeneracy. This leads to Hund's-like rule, where within each shell the total spin is maximized. As a result, we obtain an oscillating net spin: every time the number of electrons is a multiple of two, we obtain $S = 1$. When $N = 8$, the system can neither be considered thin nor wide enough to have strongly degenerate edge states. The competition between Hund's rules for double shells and net edge magnetization give rise to rather complex oscillations of the total spin as a function of the number of electrons.

In Fig. 6.30 we show the configuration interaction results for the same ribbon as in Fig. 6.29, but in the Möbius configuration. Although both ribbons have the same number of atoms, unlike in the cyclic case, the charge half-filling of the degenerate band in the Möbius configuration occurs at $N_e = 7$, and not at the charge neutrality point, $N_e = 8$. For the wide ribbon ($N = 14$), at half-filling $N_e = 7$, the system is ferromagnetic. This can be understood from the fact that the Möbius configuration is a one-edged system, behaving like the zigzag edge of semi-infinite bulk graphene, or of a triangular graphene quantum dot [2, 3, 26]. The magnetization is lost in the charge-neutral case. This is in agreement with the earlier mean-field Hubbard calculations using s-type orbitals, where opposite edges are found to be in an antiferromagnetic configuration, but with a spin domain wall helping to overcome the magnetic frustration along the zigzag edge of the Möbius strip [36, 37]. Away from charge neutrality, for the cyclic ribbon, the degeneracy between edge states is lifted, and the shell structure becomes prominent. There is a difference, however, due to the broken electron-hole symmetry in the single-particle energy spectrum of the Möbius ribbon. Thus, the Hund's rule which maximizes the total spin within a shell still applies, but the total spin spectrum does not have the electron-hole symmetry anymore.

References

1. E.H. Lieb, Phys. Rev. Lett. **62**, 1201 (1989)
2. A.D. Güçlü, P. Potasz, O. Voznyy, M. Korkusinski, P. Hawrylak, Phys. Rev. Lett. **103**, 246805 (2009)
3. P. Potasz, A.D. Güçlü, A. Wójs, P. Hawrylak, Phys. Rev. B **85**, 075431 (2012)
4. B. Wunsch, T. Stauber, F. Guinea, Phys. Rev. B **77**, 035316 (2008)
5. I. Romanovsky, Y. Yannouleas, U. Landman, Phys. Rev. B **79**, 075311 (2009)
6. P. Hawrylak, A. Wójs, J.A. Brum, Phys. Rev. B **55**, 11397 (1996)
7. M. Ezawa, Phys. Rev. B **76**, 245415 (2007)
8. J. Fernandez-Rossier, J.J. Palacios, Phys. Rev. Lett. **99**, 177204 (2007)
9. W.L. Wang, S. Meng, E. Kaxiras, Nano Lett. **8**, 241 (2008)
10. S.L. Sondhi, A. Karlhede, S.A. Kivelson, E.H. Rezayi, Phys. Rev. B **47**, 16419 (1993)
11. A.H. MacDonald, H.A. Fertig, L. Brey, Phys. Rev. Lett. **76**, 2153 (1996)
12. K. Nomura, A.H. MacDonald, Phys. Rev. Lett. **96**, 256602 (1996)
13. P. Plochocka, J.M. Schneider, D.K. Maude, M. Potemski, M. Rappaport, V. Umansky, I. Bar-Joseph, J.G. Groshaus, Y. Gallais, A. Pinczuk, Phys. Rev. Lett. **102**, 126806 (2009)
14. M. Ciorga, A.S. Sachrajda, P. Hawrylak, C. Gould, P. Zawadzki, S. Jullian, Y. Feng, Z. Wasilewski, Phys. Rev. B **61**, R16315 (2000)
15. A.D. Güçlü, Q.F. Sun, H. Guo, R. Harris, Phys. Rev. B **66**, 195327 (2002)
16. O. Voznyy, A.D. Güçlü, P. Potasz, P. Hawrylak, Phys. Rev. B **83**, 165417 (2011)
17. T. Wassmann, A.P. Seitsonen, A.M. Saitta, M. Lazzeri, F. Mauri, Phys. Rev. Lett. **101**, 096402 (2008)
18. P. Koskinen, S. Malola, H. Häkkinen, Phys. Rev. B **80**, 073401 (2009)
19. Ç.Ö. Girit, J.C. Meyer, R. Erni, M.D. Rossell, C. Kisielowski, L. Yang, C. Park, M.F. Crommie, M.L. Cohen, S.G. Louie, A. Zettl, Science **323**, 1705 (2009)
20. A. Chuvilin, J.C. Meyer, G. Algara-Siller, U. Kaiser, New J. Phys. **11**, 083019 (2009)
21. J.M. Soler, E. Artacho, J.D. Gale, A. Garcia, J. Junquera, P. Ordejon, D. Sanchez-Portal, J. Phys. Condens. Matter **14**, 2745 (2002)
22. J.P. Perdew, K. Burke, M. Ernzerhof, Phys. Rev. Lett. **77**, 3865 (1996)
23. A.D. Güçlü, P. Potasz, P. Hawrylak, Phys. Rev. B **84**, 035425 (2011)
24. M. Ezawa, Phys. Rev. B **81**, 201402 (2010)
25. P. Potasz, A.D. Güçlü, P. Hawrylak, Phys. Rev. B **81**, 033403 (2010)
26. A.D. Güçlü, P. Potasz, P. Hawrylak, Phys. Rev. B **82**, 155445 (2010)
27. K.F. Mak, C.H. Lui, J. Shan, T.F. Heinz, Phys. Rev. Lett. **102**, 256405 (2009)
28. P. Potasz, A.D. Güçlü, O. Voznyy, J.A. Folk, P. Hawrylak, Phys. Rev. B **83**, 174441 (2011)

29. P. Potasz, A.D. Güçlü, P. Hawrylak, Phys. Rev. B **82**, 075425 (2010)
30. C.L. Kane, E.J. Mele, Phys. Rev. Lett. **95**, 226801 (2005)
31. M.Z. Hasan, C.L. Kane, Rev. Mod. Phys. **82**, 3045 (2010)
32. S. Konschuh, M. Gmitra, J. Fabian, Phys. Rev. B **82**, 245412 (2010)
33. A.D. Guclu, M. Grabowski, P. Hawrylak, Phys. Rev. B **87**, 035435 (2013)
34. R. Herges, Chem. Rev. **106**, 4820 (2006)
35. D. Ajami, O. Oeckler, A. Simon, R. Herges, Nature **426**, 819 (2003)
36. K. Wakabayashi, K. Harigaya, J. Phys. Soc. Jpn. **72**, 998 (2003)
37. K. Harigaya, A. Yamashiro, Y. Shimoi, K. Wakabayashi, Synth. Met. **152**, 261 (2005)
38. D. Jiang, S. Dai, J. Phys. Chem. C **112**, 5348 (2008)
39. E.W.S. Caetano, V.N. Freire, S.G. dos Santos, D.S. Galvao, F. Sato, J. Chem. Phys. **128**, 164719 (2008)
40. Z.L. Guo, Z.R. Gong, H. Dong, C.P. Sun, Phys. Rev. B **80**, 195310 (2009)
41. X. Wang, X. Zheng, M. Ni, L. Zou, Z. Zeng, Appl. Phys. Lett. **97**, 123103 (2010)

Chapter 7
Optical Properties of Graphene Nanostructures

Abstract This chapter describes the optical properties of graphene quantum dots. It discusses the size, shape and edge dependence of the energy gap, optical joint density of states, excitons, charged excitons, optical spin blockade and optical control of the magnetic moment in triangular graphene quantum dots with zigzag edges. The electronic and optical properties of colloidal graphene quantum dots, and in particular the spectrum of band-edge excitons is described.

7.1 Size, Shape and Type of Edge Dependence of the Energy Gap

In Chaps. 2–6 we have seen that the electronic and magnetic properties of the graphene quantum dot strongly depend on its size, shape, and the character of the edge. This was demonstrated by comparing the electronic properties of different graphene quantum dots, including hexagonal dots with either armchair or zigzag edges and triangular dots with armchair or zigzag edges (see Fig. 4.3). Thus, we might anticipate that the optical properties are also strongly dependent on size, shape and the edge character of the graphene quantum dot [1–3]. Indeed, finite size opens an energy gap across the Fermi level in graphene quantum dots. The energy gap corresponds to the lowest possible electronic transition from the top of the occupied valence band to the bottom of the empty conduction band, as shown in the inset in Fig. 7.1 for a hexagonal graphene quantum dot with armchair edges. The energy gap determines the lowest energy at which the quantum dot may absorb light. In Fig. 7.1 we show the dependence of the energy gap computed by exact diagonalization of the tight-binding Hamiltonian on the number of atoms N for hexagonal dots with armchair and zigzag edges and triangular dots with zigzag edges. The inset in the left lower corner of Fig. 7.1 shows the calculated energy spectrum for a $N = 114$ hexagonal quantum dot with armchair edges, redrawn from Fig. 4.3a. The double-headed red arrow indicates the energy gap E_g separating the occupied valence band states from the empty conduction band states. The energy gap measured in units of the hopping matrix element t as a function of the number of atoms N is shown in Fig. 7.1

A.D. Güçlü et al., *Graphene Quantum Dots*,
NanoScience and Technology, DOI 10.1007/978-3-662-44611-9_7

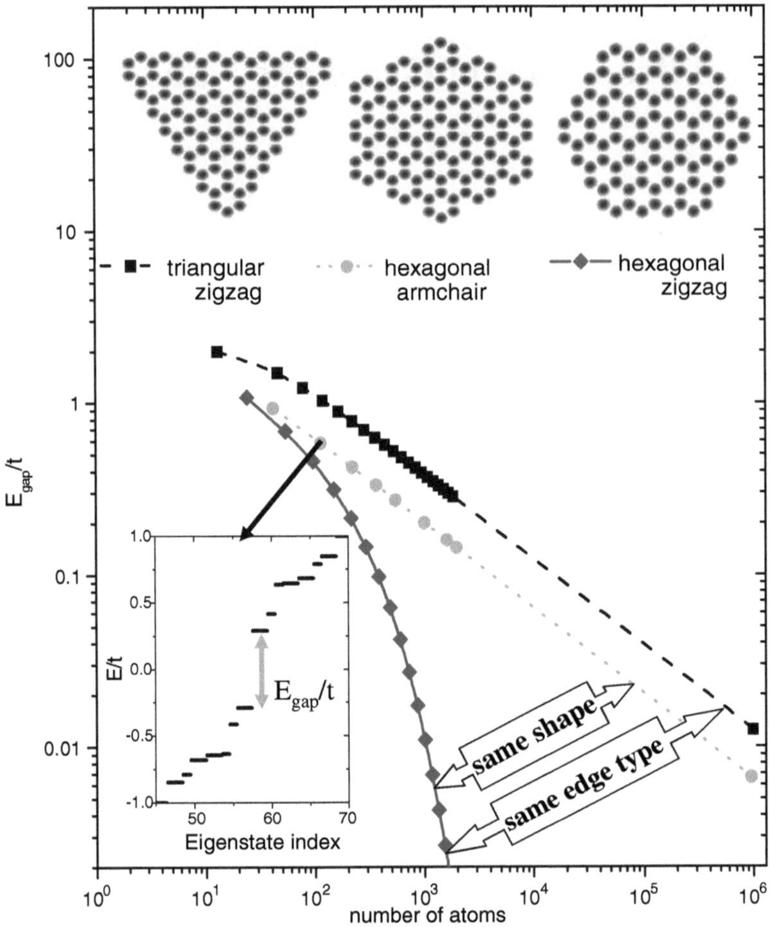

Fig. 7.1 TB bandgap energy as a function of total number of atoms N for a triangular zigzag quantum dot (*dashed line* with *squares*), hexagonal armchair quantum dot (*dotted line* with *circles*), and hexagonal zigzag quantum dot (*solid line* with *diamonds*). The inset shows the TB energy spectrum for the hexagonal armchair dot redrawn from Fig. 4.3a. Reprinted from [7]

for up to million atoms. The dotted straight line connecting computed energy values describes the gap decaying as the inverse of the square root of the number of atoms N, from hundred-to-million-atom nanostructures. This dependence is expected for confined Dirac Fermions with photon-like linear energy dispersion, $E \propto k$, where k is the wavevector of both an electron and a hole confined to a quantum dot. The wavevector $k = 2\pi/\lambda$ corresponds to wavelength which in turn must be a multiple of the diameter $2R$ of the quantum dot. Adding the energy of electron and a hole gives the energy gap $E_g \propto 1/R$. But the area of the dot, R^2, is proportional to the number of atoms N, $R^2 \propto N$, hence $E_g \propto 1/\sqrt{N}$, as pointed out in [4–7] based on calculations for hundreds of atoms. We see that the dependence of the energy gap from tens to millions of atoms follows the prediction resulting from the Dirac spectrum, dotted line, very well. Maintaining the hexagonal structure but changing the edge from armchair to zigzag has a significant effect on the evolution of the bandgap with size of the quantum dot. The energy gap of a hexagonal structure with zigzag edges decreases much faster as the number of atoms increases. This is due to the zigzag edges leading to states localized at the edge of the quantum dot, as shown in Fig. 4.4c. Let us now investigate the change in the optical gap as a function of the shape of a quantum dot. If we deform GQD with zigzag edge from hexagonal to triangular, as was shown in Chap. 2, in addition to valence and conduction bands the energy spectrum contains a shell of degenerate levels at the Fermi level. We define the bandgap in a TGQD as the energy difference between the topmost valence state, below the energy of the degenerate shell, to the lowest conduction band state above the energy of the degenerate shell. Despite the presence of zigzag edges and the zero-energy shell, the energy gap of the triangular quantum dot with zigzag edge follows the power law $E_{gap} \propto 1/\sqrt{N}$ characteristic for Dirac electrons. This suggests that the conduction and valence band states may be well described by the Dirac Fermion model. For all three shapes of quantum dots studied the energy gap varies from $E_{gap} \simeq 2.5\,\text{eV}$ corresponding to green light for a quantum dot with $N \simeq 100$ atoms and a diameter $\sim 1\,\text{nm}$ to $E_{gap} \simeq 30\,\text{meV}$ corresponding to terahertz radiation for a quantum dot with $N \simeq 10^6$ atoms and a diameter $\sim 100\,\text{nm}$. In semiconductor quantum dots the lowest energy gap is given by the bulk energy gap. Size quantization allows to increase the bandgap from the bulk value. With graphene, a semimetal with zero bandgap, the possibility of engineering the bandgap by size, shape and edge spans the energy range from terahertz to UV.

7.2 Optical Joint Density of States

We now discuss the interaction of the graphene quantum dot with photons in the electric dipole approximation. In this approximation the light-matter interaction is described by the Hamiltonian $V = \mathbf{E} \cdot \mathbf{r}$ where $\mathbf{E}$ is the electric field of a photon with energy ω and $\mathbf{r}$ is the position of an electron. The photon can be absorbed if its energy ω matches the energy of the transition from the initial occupied state $|i\rangle$ to the final unoccupied state $|f\rangle$, $\omega = E_f - E_i$. The probability amplitude for such a transition is given by $|\mathbf{E} \cdot \mathbf{d}_{if}|^2$ where $\mathbf{d}_{if} = \langle f|\mathbf{r}|i\rangle$ is the dipole moment matrix element connecting the two states. When both initial and final states are expanded in atomic orbitals localized on carbon atoms, the dipole matrix element can be expressed in terms of dipole elements involving transitions between two atoms s and s':

$$\mathbf{d}_{ss'} = D_{ss'}\hat{\mathbf{r}}_{ss'} + \mathbf{R}_s\delta_{s,s'}, \tag{7.1}$$

where $\mathbf{R}_s$ is the position of the atom s, $\hat{\mathbf{r}}_{ss'}$ is a unit vector between atoms s and with $(rss = 0)s'$, $D_{ss'} = 0.3433$ for nearest-neighbors and $D_{ss'} = 0.0873$ for next nearest neighbors in atomic units.

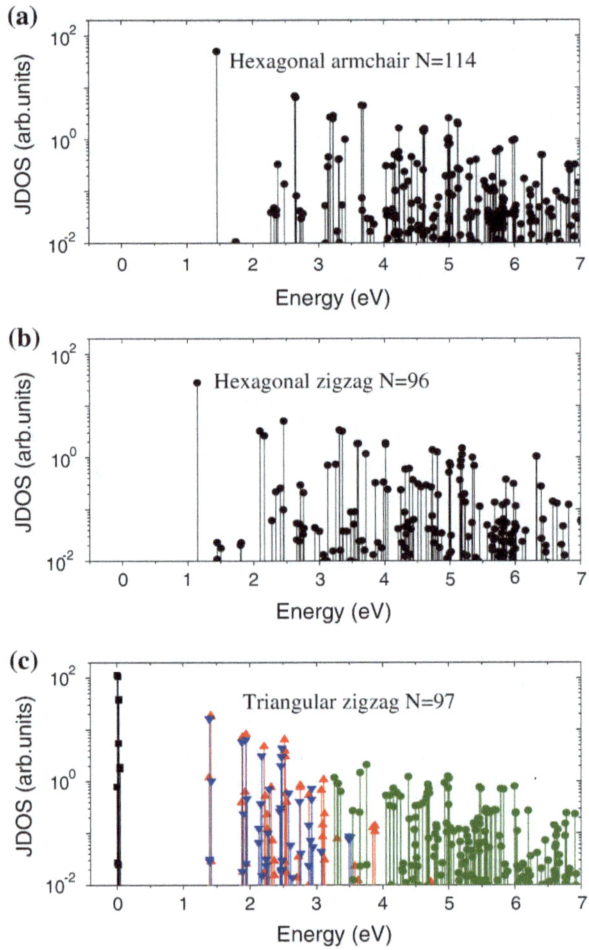

Fig. 7.2 Optical joint density of states for **a** hexagonal armchair structure with $N = 114$ atoms, **b** hexagonal zigzag structure with $N = 96$ atoms, and **c** triangular zigzag structure with $N = 97$ atoms. Due to the presence of zero-energy states in triangular zigzag structure, different classes of optical transitions exist represented by different symbols and colors. Reprinted from [7]

The joint optical density of states (JDOS) contains information about all optically active transitions from the valence to the conduction band:

$$I(\omega) = \sum_{f,i} P_i |d_{if}|^2 \delta(\omega - (E_f - E_i)). \tag{7.2}$$

where P_i is the probability that the initial state is occupied. We calculated the JDOS for the three structures with energy spectra similar to those shown in Fig. 4.3a, c and d, as a function of photon energy ω. The results are shown in Fig. 7.2. We see that the transitions corresponding to three different graphene quantum dots are optically active, including transitions at the energy gap E_g. The JDOS is modulated as some of the transitions between different energy levels have higher oscillator strength. The dipole transitions for the hexagonal armchair structure with $N = 114$, shown in Fig. 7.2a, are similar to those for the hexagonal zigzag structure shown in Fig. 7.2b. The lowest energy transition at $\omega = 1.5$ eV is lower for the structure with the zigzag edge, $\omega = 1.1$ eV, without loss of the oscillator strength. However, for the triangular zigzag structure we observe a different absorption spectrum with a group of transitions at THz energies (near $\omega = 0$) which is absent in the two other structures. These photon energies correspond to transitions between the states of the zero-energy shell. The inclusion of second neighbor hopping removes the degeneracy and allows for the intra-shell transitions at THz energies. In addition to intra-shell transitions, the on i set of valence to conduction band transitions is at significantly higher energy $\omega = 3$ eV. However, transitions from VB into the zero-energy states and out of zero-energy states into the CB start at 1.5 eV, i.e., in the middle of the valence to conduction band gap. There triangular graphene quantum dots appear to be good candidates for the intermediate band solar cells.

7.3 Triangular Graphene Quantum Dots With Zigzag Edges

7.3.1 Excitons in Graphene Quantum Dots

The JDOS discussed in the previous section does not include the effects of electron-electron interactions. In order to take into account the electron-electron interactions in the ground state and in the excited states, we introduce the Hamiltonian for electrons above the Fermi level and holes created below the Fermi level of an interacting graphene quantum dot:

$$\begin{aligned} H = & \sum_{s'\sigma} \varepsilon_{s'\sigma} a^{\dagger}_{s'\sigma} a_{s'\sigma} + \sum_{s\sigma} \varepsilon_{s\sigma} h^{\dagger}_{s\sigma} h_{s\sigma} \\ & + \frac{1}{2} \sum_{\substack{s',p',d',f',\\ \sigma\sigma'}} \langle s'p'|V|d'f'\rangle a^{\dagger}_{s'\sigma} a^{\dagger}_{p'\sigma'} a_{d'\sigma'} a_{f'\sigma} + \frac{1}{2} \sum_{\substack{s,p,d,f,\\ \sigma\sigma'}} \langle sp|V|df\rangle h^{\dagger}_{s\sigma} h^{\dagger}_{p\sigma'} h_{d\sigma'} h_{f\sigma} \\ & + \sum_{\substack{s',p,d,f',\\ \sigma\sigma'}} \left(\langle ds'|V_{ee}|f'p\rangle - (1-\delta_{\sigma\sigma'})\langle ds'|V_{ee}|pf'\rangle\right) a^{\dagger}_{s\sigma} h^{\dagger}_{p\sigma'} h_{d\sigma'} a_{f'\sigma}, \end{aligned} \tag{7.3}$$

where indices (s, p, d, f) correspond to states below the Fermi level, and (s', p', d', f') correspond to states above the Fermi level. Operators $h^{\dagger}_{s\sigma}$ $(h_{s\sigma})$ create (annihilate) a hole in the valence band of quasiparticles obtained using a combination of tight-binding and the Hartree-Fock approach (TB+HF). Terms in the first line correspond to kinetic energies of electrons and holes. Terms in the second line correspond to interactions between electrons (the first term) and interactions between holes (the second term). Terms in the third line describe attractive direct interaction (the first term) and repulsive exchange interaction (the second term) between an electron and a hole. This Hamiltonian describes a filled valence band obtained in the TB+HF approximation and additional electrons filling up the degenerate zero-energy shell. The number of additional electrons is controlled by the external gate. When the number of additional electrons equals the number of states in the degenerate shell, the TGQD is charge-neutral. This is the Hamiltonian studied in Chap. 6. In addition to electrons controlled by the gate we now added photoexcited electrons and photoexcited holes created by interaction with light. The attractive electron-hole interaction describes excitonic effects where the photoexcited electron interacts with a valence hole but is indistinguishable from the electrons of the degenerate shell. The situation is analogous to the description of the optical properties of interacting electrons on the Haldane sphere [8].

The interaction of the TGQD with light is described by the excitonic absorption spectrum (exciton spectral function) $A(\omega)$ involving transitions between the N-electron ground state $|GS\rangle$ and final excited $N+1+h$ states $|f\rangle$ with a photoexcited electron and a valence hole:

$$A(\omega) = \sum_f |\langle f|P^\dagger|GS\rangle|^2 \delta(\omega - (E_f - E_{GS})), \tag{7.4}$$

where $P^\dagger = \sum_{ss'} \delta_{\sigma\bar{\sigma}'} d_{s',s} h^\dagger_{s\sigma} a^\dagger_{s'\sigma'}$ is the polarization operator. $P^\dagger$ adds an exciton to the ground states of the TGQD with simultaneous annihilation of a photon.

We now illustrate the optical properties of triangular graphene quantum dots on the example of a TGQD with $N = 97$ atoms, for which exact many-body calculations can be carried out. For the charge-neutral case, all states of the valence band are doubly occupied with spin-up and down electrons while each state of the zero-energy shell is singly occupied with all electrons having parallel spin, which was shown in Sect. 4.1. With half-filled zero energy shell we can classify allowed optical transitions into four classes, shown in Fig. 7.3a: (i) from valence band to zero-energy degenerate band (VZ transitions, blue color); (ii) from zero-energy band to conduction band (ZC transitions, red color); (iii) from valence band to conduction band (VC transitions, green color); and finally, (iv) within zero-energy states (ZZ transitions, black color). As a consequence, there are three different photon energy scales involved in the absorption spectrum. VC transitions (green) occur above the full bandgap (2.8 eV), VZ (blue) and ZC (red) transitions occur starting at half band gap (1.4 eV), and ZZ (black) transitions occur at terahertz energies. The energies corresponding to ZZ transitions are controlled by the second-nearest-neighbor tunneling matrix element t' and by electron-electron interactions. Figures 7.3b–d illustrate in detail the effect of electron-electron and final-state (excitonic) interactions on the absorption spectra. Figure 7.3b shows the detailed VZ absorption spectrum for noninteracting electrons. This spectrum corresponds to transitions from the filled valence band to half filled shell of $N_{deg} = 7$ zero-energy states. Half filling implies that transition to each state of the zero-energy band is optically allowed. According to electronic densities of the degenerate states shown in Fig. 4.19, among the $N_{deg} = 7$ zero-energy states,

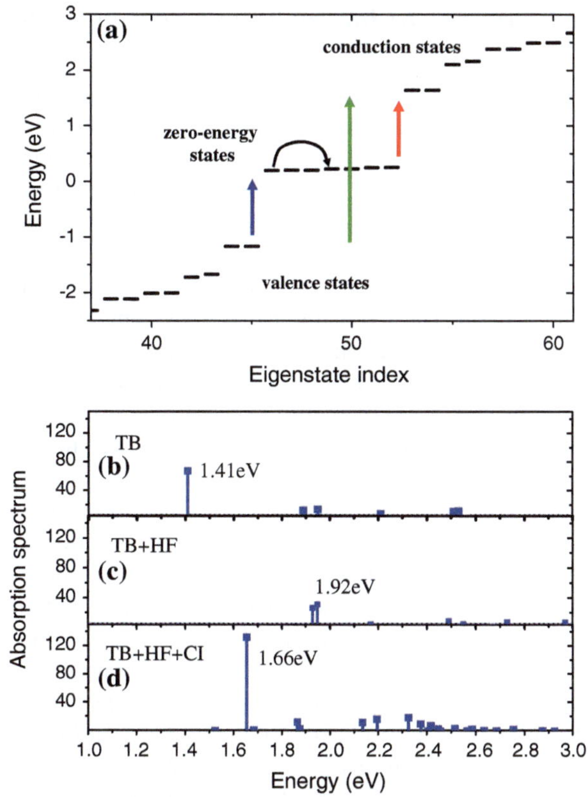

Fig. 7.3 **a** Possible optical transitions in TGQD consisting of $N = 97$ atoms. The *colored arrows* represent optical transitions from VC *green*, VZ *blue*, ZC *red*, and ZZ *black*. **b–d** Shows the effect of electron-electron interactions on the VZ transitions within **c** Hartree-Fock approximation, and including **d** correlations and excitonic effect obtained from exact configuration interaction calculations. Reprinted from [7]

there are two bulk-like states, Fig. 4.19d, e, which are dipole coupled to the states at the top of the valence band resulting in the final oscillator strength of the main transition at $E = 1.41$ eV. When the electron-electron interactions are turned on at the Hartree-Fock level, the photon energies corresponding to optical transitions, $\omega = (E_f + \sum_f) - (E_i + \sum_i)$, are renormalized by the difference in quasiparticle self-energies $\sum_f - \sum_i$. The absorption spectrum, shown in Fig. 7.3c, is renormalized with transition energies blue-shifted by 0.51 eV to $E = 1.92$ eV. Finally, when final state interactions between all interacting quasielectrons and quasiholes are taken into account, the excitonic spectrum is again renormalized from the quasiparticle spectrum, with transitions red shifted from quasiparticle transitions at $E = 1.92$ eV, down to $E = 1.66$ eV. As we can see, electron-electron interactions play an important role in determining the energies and the form of the absorption spectrum, with a net blue-shift from the noninteracting spectrum by 0.25 eV for this particular quantum dot.

7.3.2 Charged Excitons in Interacting Charged Quantum Dots

We now turn to the analysis of the effect of gate tunable carrier density on the optical properties of graphene quantum dots. The finite carrier density, controlled by either a metallic gate or via doping (intercalation), has been shown to significantly modify the optical properties of graphene [9, 11–13] and semiconductor quantum dots [10].

For a triangular graphene quantum dot, the metallic gate, shown in Fig. 7.4a, changes the number of electrons in the degenerate shell from $N_{deg} = 7$ to $N_{deg} + \Delta N_{deg}$. This is illustrated in Fig. 7.4b where four electrons were removed and three electrons remain. These remaining electrons populate the degenerate shell and their properties are entirely controlled by their interactions. Alternatively, the removal of electrons from a charge-neutral shell corresponds to addition of holes. As is clear from Fig. 7.4b, such a removal of electrons allows intra-shell transitions ZZ, enhances VZ transitions by increasing the number of allowed final states and weakens the ZC transitions by decreasing the number of occupied initial states. Figure 7.4c illustrates the overall effects in the computed excitonic absorption spectra for VZ, ZC, and ZZ transitions as a function of the number of additional electrons N_{deg}. At $\Delta N_{deg} = -7$ (hole-filling factor $\nu_h = 1$), the shell is empty and VZ transitions describe an exciton built of a hole in the valence band and an electron in the degenerate shell. The absorption spectrum has been described in Fig. 7.3b–d and is composed of one main excitonic peak at 1.66 eV. There are no ZC transitions and no ZZ transitions in the terahertz range. When we populate the shell with electrons, the VZ excitonic transition turns into a band of red-shifted transitions corresponding to an exciton interacting with additional electrons, in analogy to optical processes in the fractional quantum Hall effect [8] and charged semiconductor quantum dots [10]. As the shell filling increases, the number of available states decreases and the VZ transitions are quenched while ZC and ZZ transitions are enhanced. These results show that the absorption spectrum can be tuned by shell filling, which can be experimentally controlled by applying a gate voltage. This is particularly true for the ZZ transitions in the terahertz range, which can be turned off by either emptying/filling the shell, $\Delta N_{deg} = \pm 7$, or at half filling. At half filling, electron exchange leads to spin polarization, with each state of the shell filled by a spin-polarized electron. Since photons do not flip electron spin, no intra-shell transitions are allowed and the magnetic moment of the graphene quantum dot is directly reflected in the ZZ absorption spectrum.

7.3.3 Terahertz Spectroscopy of Degenerate Shell

In Fig. 7.5 we describe the transitions for $\Delta N_{deg} = 0, \pm 1$ in detail. Figure 7.5b shows the lack of absorption for the half-filled spin-polarized shell. The right hand side illustrates the fact that photons pass through the graphene quantum dot since they are not able to induce electronic transitions and be absorbed. For $\Delta N_{deg} = -1$,

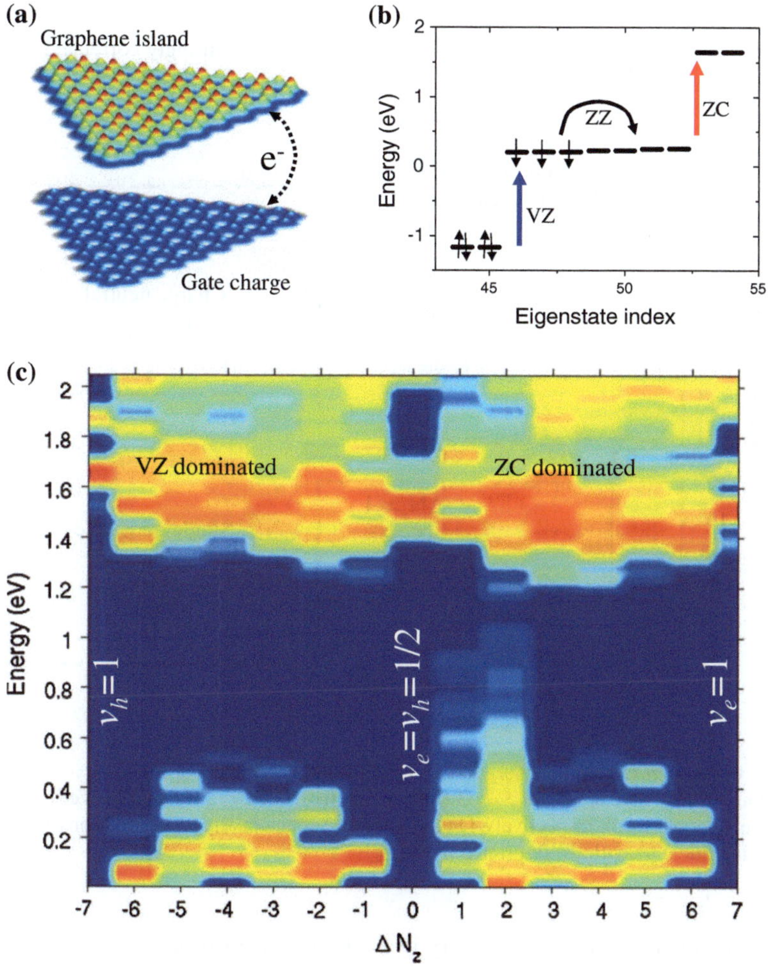

Fig. 7.4 **a** Schematic representation of TGQD with $N = 97$ carbon atoms with four electrons moved to the metallic gate. **b** Corresponding single particle TB configuration near the Fermi level. **c** Excitonic absorption spectrum in arbitrary units as a function of energy and charging ΔN_{deg}. For convenience, transitions are artificially broadened by 0.02 eV. Peaks below 0.6 eV are due to ZZ transitions, peaks above 1.2 eV are due to VZ and ZC transitions. Charge neutral case corresponds to $\Delta N_{deg} = 0$ (filling factors $\nu_e = \nu_h = 1/2$). Reprinted from [7]

Fig. 7.5c, one electron is removed creating a hole in the spin-polarized shell. Thus, the absorption spectrum corresponds to transitions from the ground state to optically allowed excited states of the hole. The absorption spectrum for an additional electron, $\Delta N_{deg} = +1$, shown in Fig. 7.5a, is dramatically different. The addition (but not subtraction) of an electron depolarizes the spins of all electrons present, with total spin of the ground state $S = 0$. The strongly correlated ground state has many configurations, which effectively allow for many transitions of the spin-up and

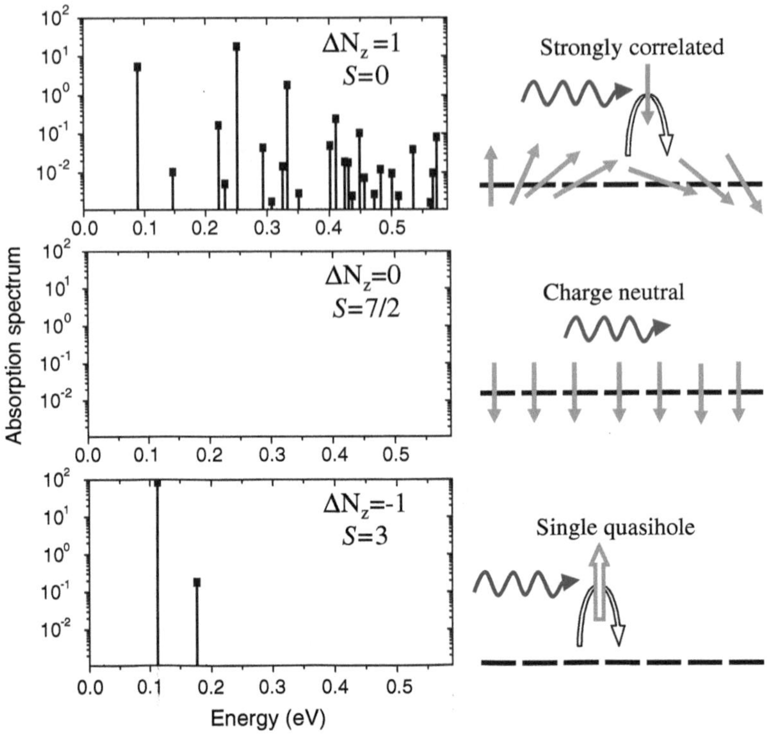

Fig. 7.5 On the *left* excitonic absorption spectrum in arbitrary units at $\Delta N_{deg} = -1, 0, 1$. Corresponding ground state spins are $S = 3$ (fully polarized), $S = 7/2$ (fully polarized), and $S = 0$ (completely depolarized), respectively. On the *right* the schematically representation of the physics involved in optical transitions. Reprinted from [7]

spin-down electrons. This asymmetry in the terahertz absorption spectra allows for the optical detection of charge of the quantum dot and correlated electron states in the degenerate electronic shell.

7.4 Optical Spin Blockade and Optical Control of Magnetic Moment in Graphene Quantum Dots

We will now show that the magnetization of triangular graphene quantum dots with zigzag edges can be manipulated optically. Indeed, while in a doped TGQDs quantum dot depolarization occurs due to electron-electron interactions, the magnetization can be recovered by absorption of a photon due to electron-hole interactions. The conversion of the photon to a magnetic moment results in a many-body effect, the optical spin blockade. The effect demonstrated here can potentially lead to efficient spin to photon conversion, quantum memories and single photon detectors [14].

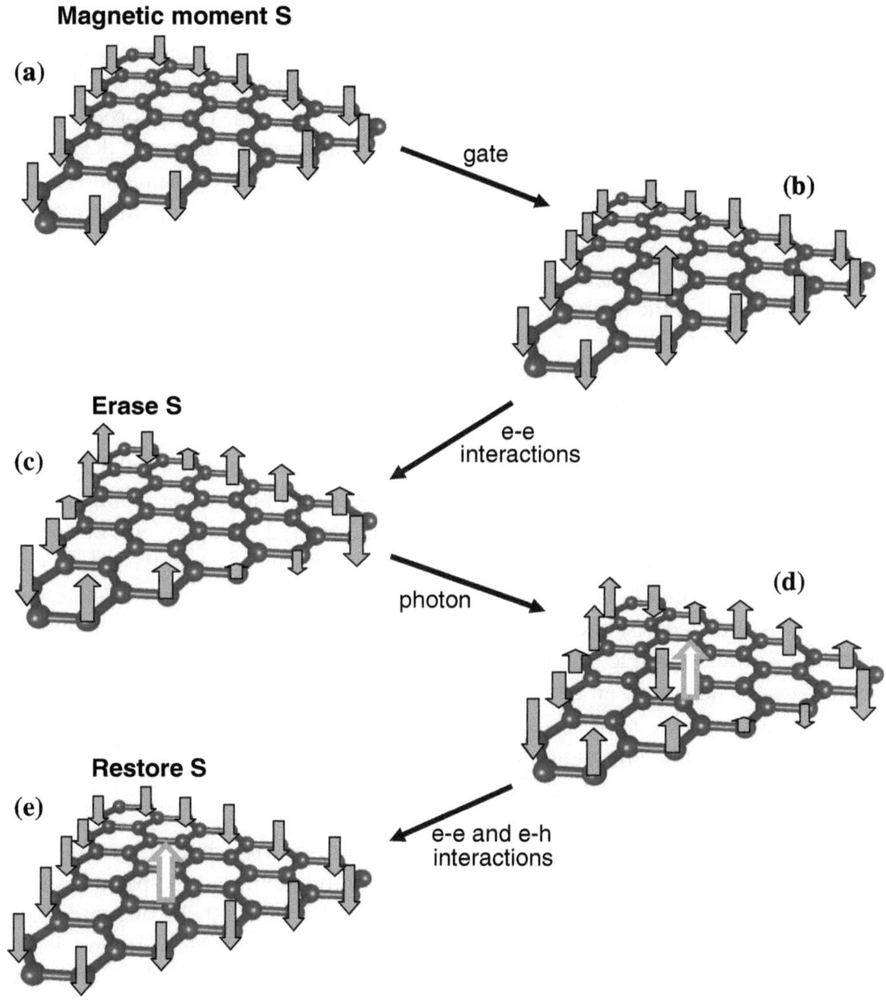

Fig. 7.6 Schematic illustration of optical control of magnetization and origin of optical spin blockade: **a** Creation of magnetic moment S; **b–c** erasure of S with addition of a single electron which destroys S; **d–e** restoration of a single photon that creates an exciton which restores magnetic moment S through e-e and e-h interactions. Reprinted from [14]

Before discussing the details of calculations and results, in Fig. 7.6 we schematically summarize the process of optical manipulation of the magnetic moment S, total spin, in a TGQD with zigzag edges. The blue balls illustrate carbon atoms held together by sp^2 bonds, and red arrows illustrate p_z electron spin density. When the TGQD is charge neutral, Fig. 7.6a, the electrons in the vicinity of zigzag edges align their spin through exchange interaction, giving rise to a net magnetic moment S. If the TGQD is charged with a single additional electron by a gate below a TGQD, the added electron must have spin opposite to the magnetization S (Fig. 7.6b). Through electron-electron interactions, electrons attempt to align their spin with the added electron, inducing spin depolarization as illustrated in Fig. 7.6c. However, the spin polarization can be recovered by absorption of a single photon. The absorbed photon creates a hole in the valence band (thick arrow) and an electron in the degenerate shell at the zero-energy Fermi level as shown in Fig. 7.6d. The exchange interaction between the valence hole and all the electrons in the degenerate shell aligns the spin of electrons in the degenerate shell and restores the magnetic moment (Fig. 7.6e). Hence one can erase the magnetic moment with a gate and restore it optically making it possible to control the magnetization of a graphene quantum dot with zigzag edges through optical spin blockade.

Figure 7.7a–b summarizes the depolarization process discussed in Sect. 6.1. The left panel of Fig. 7.7a shows the single-particle energy levels of the noninteracting TGQD. The arrows schematically show a single configuration of $N_e = 7$ quasi-electrons with all electron spins aligned. The total spin S of this spin-polarized configuration is $S = 7/2$ as we have seen before. There are many other configurations possible with total spin varying from $S = 7/2$ to $S = 1/2$. The low-energy spectra for the charge-neutral TGQD for different possible total spin S are shown in Fig. 7.7a, right panel. The ground state corresponds to a maximally spin polarized state with $S = 3.5$, as indicated by the circle. Figure 7.7b shows the effect of the additional electron on single-particle (left) and many-particle (right) spectrum of TGQD. In a single-particle spectrum, an additional electron is added to the spin-polarized configuration, also shown in Fig. 7.7b. This electron has a spin opposite to the total spin of the TGQD. Such configuration has a total spin of $S = 7/2 - 1/2 = 3$. Fig. 7.7b, right panel, shows the low-energy spectrum of the interacting system. The ground state, marked with a circle, has total spin $S = 0$.

Figure 7.7c shows the new effect of absorption of a single photon in a charged TGQD of Fig. 7.7b. In the left panel we show the noninteracting single-particle states. The photoexcited configuration consists of a spin-polarized shell, one additional electron with opposite spin and a photo-excited opposite-spin electron and a hole in the valence band, i.e., an exciton X. The right panel of Fig. 7.7c shows the low-energy spectrum of the interacting electron-hole system. We see that the ground state corresponds to total spin $S = 6/2$. Since the optically excited exciton X is in a singlet state, i.e., does not carry net spin, the ground state total spin $S = 6/2$ corresponds to a configuration shown in Fig. 7.7b and left panel of Fig. 7.7b. Hence, the addition of an exciton to the charged TGQD restored the maximally polarized state. We can understand this remarkable effect as follows. When the system is photoexcited, a valence electron is transferred into the zero-energy shell leaving a hole behind. The addition of an extra electron to the strongly correlated spin $S = 0$ state does not change the spin polarization, resulting in a $S = 1/2$ spin depolarized ground state. However, if this additional electron is accompanied by the valence hole, a significant rearrangement of electronic correlations takes place. The introduction of the valence hole spin maximizes the exchange energy between the valence hole and electrons in a degenerate zero-energy shell only if they have aligned spins. Hence there is a competition between electronic correlations in the charged degenerate shell which

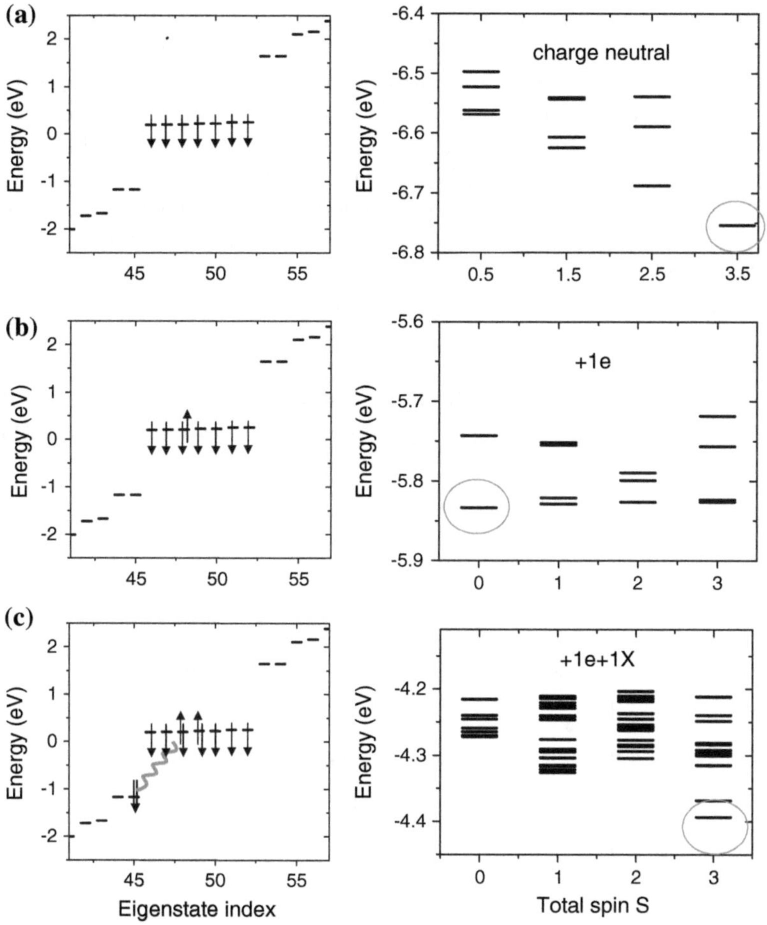

Fig. 7.7 Noninteracting *left panels* and many-body *right panels* energy spectra showing the ground state total spin of **a** charge neutral, **b** charged, and **c** charged and photoexcited quantum dot with seven zero-energy states. Reprinted from [14]

destroy spin polarization and exchange interaction with the valence hole which favor spin polarized state. Exact diagonalization of the interacting electron system shows that the exchange with the valence hole wins and, as a result, for optically excited system, the total spin is maximized: the electron total spin is $S_e = |N_d - 2|/2$ due to the two extra spins in the zero energy shell. Since the valence hole total spin is $S_h = -1/2$, the net spin of the system is given by $S = |N_d - 1|/2$ ($S = 3$ in our example).

The maximal spin polarization of the photo-excited TGQD is observed not only at filling factor $\nu = 1$ but at all filling factors. Figure 7.8 shows the calculated ground state total electronic spin S_e of TGQD as a function of the number of electrons (top) and filling fraction ν of the zero-energy shell. The black curve shows the total spin of the initial state and the red curve shows the total spin after absorption of a photon, i.e., with the exciton X. Without exciton, away from the charge neutrality, depolarization occurs for one added electron, $\nu = (N_d+1)/N_d = 8/7$, and for two added electrons, $\nu = (N_d + 2)/N_d = 9/7$. By contrast, the zero-energy shell after illumination is spin-polarized at all filling factors. Blue and red arrows show the difference between the total spin of the initial and final, photoexcited, states. The blue arrow corresponds to the spin difference equal to a single electron spin while the red arrow points to a larger difference.

As we demonstrate below, the large spin difference between the initial and final states, shown by red dashed arrows in Fig. 7.8, causes an optical spin blockade in absorption and emission spectra.

As explained in Sect. 7.3.1 the absorption spectrum is related to the spectral function $A(\omega)$ describing annihilation of a photon and addition of exciton to a TGQD

$$A(\omega) = \sum_f |\langle M^f|P^\dagger|M^i\rangle|^2\delta(\omega - (E_f - E_i)), \tag{7.5}$$

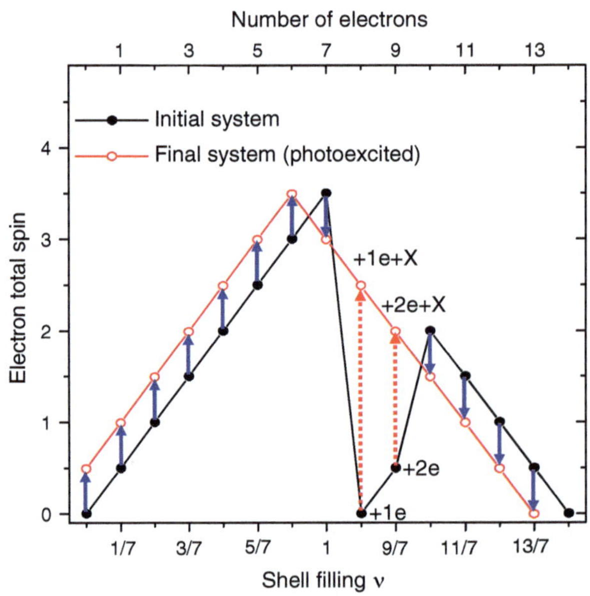

Fig. 7.8 Ground state total spin as a function of filling of the zero energy states of the system described in Fig. 7.7, with and without optical activation. Magnetization of the system is stabilized by the presence of an exciton. Optically allowed and blockaded transitions are shown with *blue* and *red arrows* respectively. Reprinted from[14]

which involves transitions between the initial many-body state $|M^i\rangle$ and all final states $|M^f\rangle$ connected by the polarization operator $P^\dagger = \sum \delta_{\sigma\bar{\sigma}'}\langle p|\mathbf{r}|q\rangle b^\dagger_{p\sigma'} h^\dagger_{q\sigma}$ creating an electron in the zero-energy shell and a hole in the valence band. The many-body matrix element contains a term $< f, N_e+1, S_e^f | b^\dagger_{p\sigma} | S_e^i, N_e, i >$ in which an electron with spin $\sigma = \pm 1/2$ in a single particle state p is added to N_e electrons in the initial many-body state i with total spin S_e^i. The resulting $N_e + 1$ state with spin $S = S_e^i \pm 1/2$ must have a finite overlap with the final state with total spin S_e^f. The overlap is finite if the total spin difference between the initial and final many-body states equals the spin of one added electron. The computed spin difference between the initial and final states in the absorption process is shown with arrows in Fig. 7.8. Blue arrows correspond to allowed transitions with spin difference of 1/2, while blocked transitions are shown as red arrows.

7.5 Optical Properties of Colloidal Graphene Quantum Dots

Recently, colloidal, solution processable graphene quantum dots (CGQDs) with well-defined structure were fabricated [15–19] and absorption and emission of solutions containing CGQDs were measured [17]. Two classes of dots, one with $N = 168$ and one with $N = 132$ carbon atoms, illustrated in Fig. 7.9, were obtained.

The number of atoms in each dot was determined from mass spectrometry, while the symmetry was inferred through the solution chemistry and infrared vibrational spectra. Since the CGQDs are suspended in solution, whose dielectric constant can be tuned, their optical response can be studied as a function of their size and shape, as well as the strength of the Coulomb interactions. Indeed, optical absorption spectra reveal a clear dependence of the position of the absorption edge on the number of atoms [15–18]. The fluorescence and phosphorescence spectroscopy [17] shows the existence of a gap between emission and absorption spectra interpreted in terms of the energy difference between the singlet and triplet exciton states.

In Chap. 5 we have studied the effects of interactions in such CGQD with $N = 168$. In this section we will investigate optical properties of the $N = 168$ CGQD in detail and compare the results with experiment [20].

7.5.1 Optical Selection Rules for Triangular Graphene Quantum Dots

The triangular $N = 168$ CGQD is rotationally symmetric and exhibits all point symmetries of the graphene sheet. If we start with an atom A in a CGQD, we can find atoms B and C which form the corners of an equilateral triangle. We now transform the three p_z orbitals (ϕ_A, ϕ_B, ϕ_C) into their linear combinations

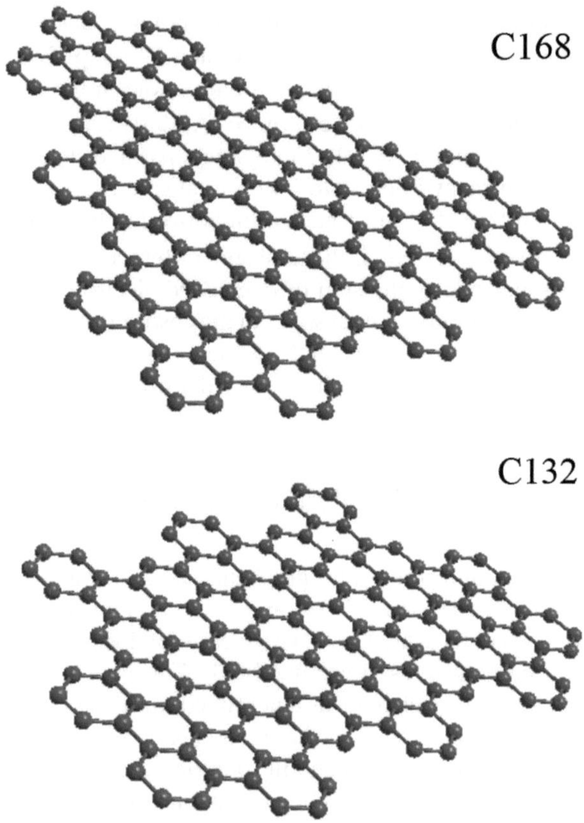

Fig. 7.9 Graphene quantum dots with 168 and 132 atoms. C168 exhibits all point symmetries of the graphene sheet

$$\phi_0 = \frac{1}{\sqrt{3}}(\phi_A + \phi_B + \phi_C) \tag{7.6}$$
$$\phi_{+1} = \frac{1}{\sqrt{3}}(\phi_A + e^{i(2\pi/3)\times 1}\phi_B + e^{i(2\pi/3)\times 2}\phi_C)$$
$$\phi_{-1} = \frac{1}{\sqrt{3}}(\phi_A + e^{-i(2\pi/3)\times 1}\phi_B + e^{-i(2\pi/3)\times 2}\phi_C).$$

Rotating the single-particle basis transforms the tight-binding Hamiltonian into a block-diagonal form with subspaces characterized by the quantum number $m = \{0, +1, -1\}$ or $m = \{0, 1, 2\}$ with index "m" appearing in exponents in Eq.7.6 [20].

We now relate the triangular symmetry to the dipole elements and optical selection rules. Expanding the rotationally invariant eigenvectors in the subspace m, $|\nu, m\rangle$,

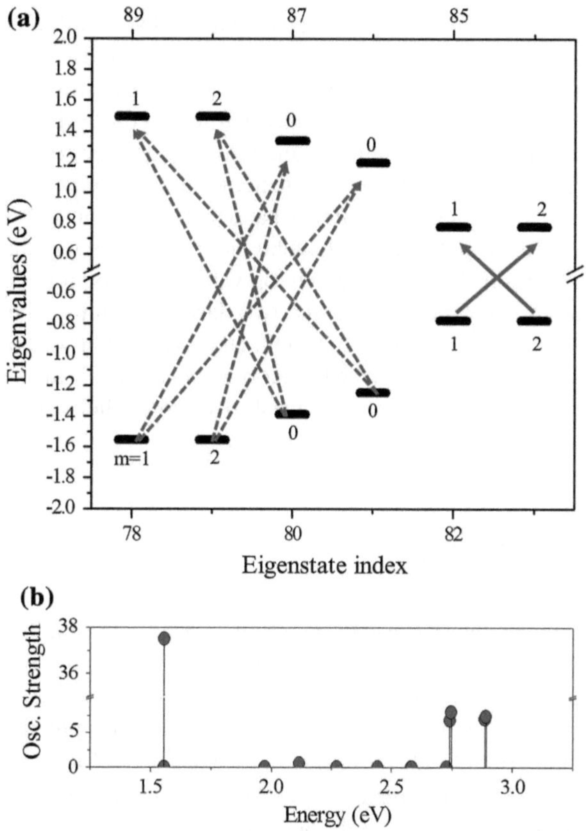

Fig. 7.10 **a** Tight-binding energy levels for C168 for $t = -3.0\,\text{eV}$, $t_2 = -0.1\,\text{eV}$ and $\kappa = 5$. Only several levels around the Fermi level are shown. *Dashed lines* between CB and VB levels indicate a weak oscillator strength while the *solid line* corresponds to strong transitions. **b** Oscillator strengths of transitions within the window of 6 CB and 6 VB levels. The strongest line around 1.5 eV corresponds to a transition between the degenerate CBM and VBM levels, while the second set of transitions at around 2.75 and 2.9 eV are due to transitions between the higher lying m = 0 and m = 1, 2 levels. Reprinted from [20]

in terms of localized orbitals, and assuming circular polarization of light $\varepsilon_{\pm}$, after lengthy algebra, we find that the dipole elements between conduction band and valence band energy levels satisfy the selection rule:

$$\langle \nu', m' | \boldsymbol{\varepsilon} \cdot \mathbf{r} | \nu, m \rangle = \delta_{m',m\pm1} C_{m,m',\nu,\nu'}, \tag{7.7}$$

where C is a constant determined numerically.

Arrows in Fig. 7.10a show the optical transitions with a finite matrix element while Fig. 7.10b shows all possible transition energies along with their dipole strength between the highest (lowest) three valence band (conduction band) states. We see indeed that the selection rule $\delta_{m',m\pm1}$ is satisfied and all vertical transitions $\Delta m = 0$ are dark. The transitions with $\Delta m = 1$, i.e. $0 \leftarrow 1$, $1 \leftarrow 2$, and $2 \leftarrow 0$, correspond to circularly polarized light with $\sigma = 1$ polarization, and the transitions with $\Delta m = -1$, i.e. $2 \leftarrow 1$, $1 \leftarrow 0$, and $0 \leftarrow -1$, correspond to circularly polarized light with $\sigma = -1$ polarization. We note that the degenerate exciton spectrum is also present in semiconductor quantum dots, where the degeneracy arises from the s-character of the conduction band, p-character of the valence band and strong spin-orbit coupling [21]. Figure 7.10b shows the dipole matrix elements as a function of transition energy. The lowest-energy transitions between the two top valence and bottom conduction band states correspond to two dipole-active and two dark transitions. The lowest-energy shell is separated by a gap from the next shell. However, the lack of dipole moments for some of the transitions between the higher lying $m = \{1, 2\}$ states with the $m = 0$ levels visible in Fig. 7.10b is due to the weak overlap of the wavefunctions and is unrelated to the symmetry.

7.5.2 *Band-edge Exciton*

Let us now describe the characteristic spectrum of band-edge excitons built of electron and a hole on the lowest-energy shell. We re-label the two topmost valence band states as $\{v1, v2\}$ and two lowest-energy conduction band states as $\{c1, c2\}$.

We start by filling up all the VB tight-binding orbitals with spin up/down electrons and forming the HF ground state $|HF_{gs}\rangle$ as shown in Fig. 7.11a. Next, the excitations $|p, q\rangle = b^{\dagger}_{p\uparrow} b_{q\uparrow} |HF_{gs}\rangle$ are created. The $\Delta m = \pm1$ optically active excitations are shown in Fig. 7.11b. There is only one electron-hole pair with $\Delta m = +1$ and one with $\Delta m = -1$ for a given spin of the excited electron. The energy of each pair, $E_{p,q} = \varepsilon_p - \varepsilon_q + \Sigma(p) - \Sigma(q) - \langle pq|V_{HF}|qp\rangle$, is given by a difference in single-particle energies and self-energies Σ of the electron and the hole and by the electron-hole attraction.

With two possible $\Delta m = \pm1$ states and two possible spin directions there are 4 exciton states, as shown on the right hand side of Fig. 7.12a. There is one singlet and one triplet state with $S_z = 0$ for each $\Delta m = \pm1$ state, given as $|p, q, S/T\rangle = \frac{(b^{\dagger}_{p\uparrow} b_{q\uparrow} \pm b^{\dagger}_{p\downarrow} b_{q\downarrow})}{\sqrt{2}} |HF_{gs}\rangle$. The energy of the singlet and triplet, $E_{p,q,S/T} = \varepsilon_p - \varepsilon_q + \Sigma(p) - \Sigma(q) - \langle pq|V_{HF}|qp\rangle + \langle pq|V_{HF}|pq\rangle \pm \langle pq|V_{HF}|pq\rangle$ differs due to twice the exchange energy, which pushes the singlets up in energy. Similar analysis is carried out for the two $\Delta m = 0$ (Fig. 7.11c) orbitally dark configurations as shown on the left hand side of Fig. 7.12a. Two $\Delta m = 0$ configurations of each total spin component interact, and thus their energy is renormalized. The final spectrum of the band-edge excitons is shown in the middle column (Full CI) of Fig. 7.12a. We find two bright degenerate singlet exciton states and a band of two dark singlet and four dark triplet exciton states at lower energies. If we count all possible S_z configurations, the low-energy band consists of two dark singlet and twelve dark triplet states. By comparing Fig. 7.12a obtained from full HF quasiparticles and Fig. 7.12b obtained

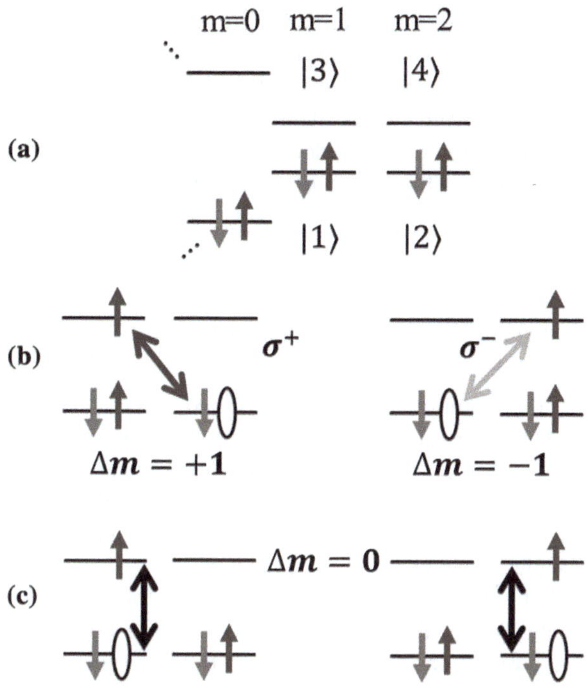

Fig. 7.11 **a** The HF ground state. **b** Single-pair excitation with total angular momentum $\Delta m = +1$ and $\Delta m = -1$. **c** Single-pair excitations from the HF ground state with total angular momentum, $\Delta m = 0$. Reprinted from [20]

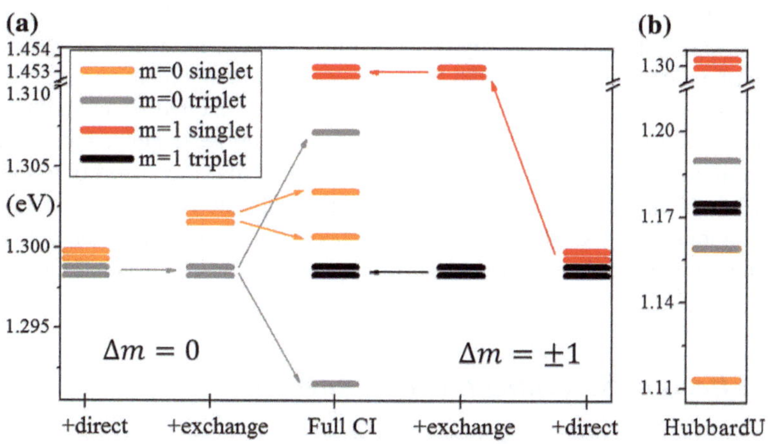

Fig. 7.12 **a** Evolution of singlet-triplet splitting with the inclusion of different interactions in C168 for $t = -3.0$ eV, $t_2 = -0.1$ eV and $\kappa = 5$ starting with the Full HF ground state. The *black lines* are $\Delta m = \pm 1$ triplets, *red* are $\Delta m = \pm 1$ singlets while $\Delta m = 0$ triplet and singlet levels are shown in *gray* and *orange*. *Left* section shows the evolution of $\Delta m = 0$ excitons while the *right* section shows the evolution of $\Delta m = \pm 1$ excitons. The *middle* section depicts all Δm levels after Full CI calculations. **b** Starting with the HubbardU ground state, the singlet-triplet splitting after full CI. Reprinted from [20]

from quasiparticles obtained by restricting Coulomb matrix elements to onsite matrix element U (HubbardU), we see that the separation of the degenerate bright singlets from the band of dark singlets and triplets is robust. However, the ordering of the levels in the dark band changes from HubbardU to full treatment of Coulomb interactions. In HubbardU approximation, dark singlet is the ground state while the inclusion of exchange interactions drives the energy of the triplet below the energy of the singlet. Hence, as expected the triplet is the lowest energy exciton state.

7.5.3 Low-Energy Absorption Spectrum

Figure 7.13 shows the evolution of the low-energy excitonic spectrum associated with the degenerate VBM/CBM states. The topmost panel shows the absorption spectrum of the noninteracting CGQD. The second panel shows the absorption in the TB+HF approximation. The self-consistent HF approach protects the rotational invariance of the $m = \{0, 1, 2\}$ subspaces but blue-shifts the energy gap due to differences in self-energy of the electron and a hole, as expected.

The third panel of Fig. 7.13 shows the band-edge exciton spectrum calculated from the HF ground state. We see that the inclusion of electron-hole attraction, exchange and electron-hole correlations red-shifts the absorption spectrum and separates in energy the singlet and triplet excitons. The two bright excitons remain degenerate, and a band of dark singlets and triplet exciton states appears at a lower energy. The last row in Fig. 7.13 shows the absorption spectrum calculated using renormalized ground and excited states obtained after the inclusion of all possible configurations with up to four pairs within the limited Hilbert space of 4 VB and 4 CB HF states. The renormalization of the energy of the ground and excited triplet states with the number of excited pairs is shown in the inset. We see that the inclusion of multi-pair excitations renormalizes both the ground state and the excited states, but does not significantly shift the transition energies nor does it remove degeneracies or change the structure of the absorption spectra. We conclude that the absorption spectrum obtained from an exciton excited out of a HF ground state is a good approximation for a semiconductor CGQD. Below we will discuss how the absorption depends on the tunneling matrix element and on screening of electron-electron interactions.

7.5.4 Effects of Screening κ and Tunneling t

The ground state properties depend strongly on the values of the strength of screening and the amplitude of the hopping term. Previous work on the ground state properties of graphene [22–24] suggest that for strong Coulomb interactions, or small values of κ, there exists a transition from a semi-metallic, weakly-interacting phase to a Mott-insulating, strongly correlated phase. Here, we discuss the phase diagram of C168 as a function of κ and t. Figure 7.14a shows the energy of the HF ground state

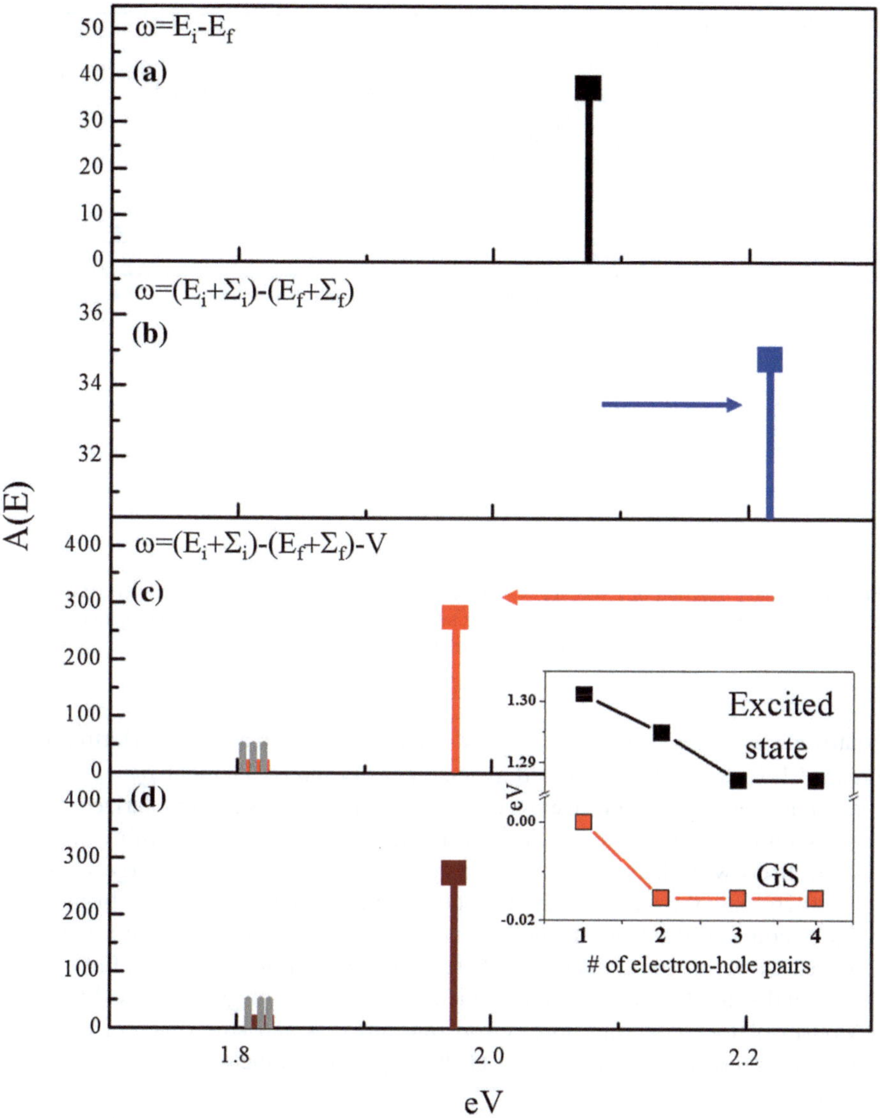

Fig. 7.13 Evolution of absorption spectrum of C168 for $t = -3.0\,\text{eV}$, $t_2 = -0.1\,\text{eV}$ and $\kappa = 5.0$, with increasing accuracy of approximations: **a** tb absorption spectrum, **b** blue-shift due to self-energy correction, **c** inclusion of electron-hole attraction and correlations and **d** renormalization of the ground state and exciton spectrum due to interaction with up to four-pair excitations. Reprinted from [20]

for the spin-polarized, $S_z = N/2$, and spin unpolarized, $S_z = 0$, states of C168 as a function of κ for $t = -4.2\,\text{eV}$. We see that, compared to the spin polarized case, the spin-unpolarized phase is the ground state for all κ down to $\kappa = 1.4$ while the spin-

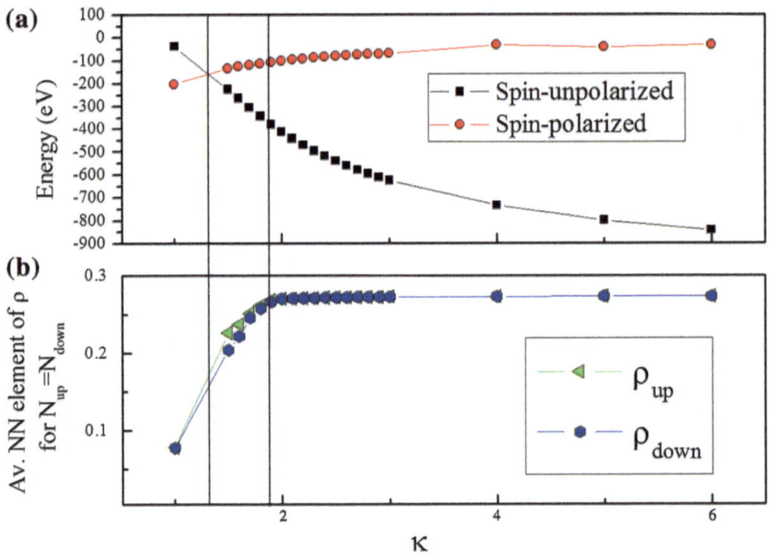

Fig. 7.14 Phase diagram of C168 at $t = -4.2$ eV, $t_2 = -0.1$ eV. **a** Ground state energy of the spin polarized and spin unpolarized C168 and **b** the nearest-neighbor density matrix element of the spin unpolarized C168 as a function of screening strength κ

polarized state is the ground state at $\kappa < 1.4$, most likely an artifact of Hartree-Fock. Figure 7.14b shows the calculated average density matrix element $\rho_\sigma = \langle c_{i\sigma}^\dagger c_{j\sigma} \rangle$ for i, j nearest neighbors, averaged over all pairs for a spin-unpolarized ground state as a function of κ. The density matrix element shows the probability of having two electrons with the same spin on nearest neighbor orbitals. For large κ we find $\rho_\sigma = 0.26$, i.e., the value for the HF state of bulk graphene [25]. The local values of ρ_σ of course differ from the bulk value at the edges even at the high-κ range. As κ decreases we see the onset of the phase transition at around $\kappa < 1.8$. For $\kappa < 1.8$ the ground state departs from the semiconducting state of graphene and becomes a Mott-insulator, with spin up electrons on lattice A and spin down electrons on lattice B. Increasing the magnitude of the hopping parameter t results in a phase transition at lower κ values.

We now discuss the evolution of the exciton spectra as a function of t and κ in the semiconducting phase. Figure 7.15a presents the results of the calculated energy of the bright degenerate singlets and $\Delta_{S/T}$ while the separation between the bright-singlet and the lowest-energy dark-triplet as a function of t and κ is given in Fig. 7.15b. We see that the energy of the bright singlets weakly depends on κ but varies with tunneling matrix element t from ~ 1 eV for $t = -2$ to 2 eV for $t = -4.2$ eV. The bright-singlet–dark-triplet separation $\Delta_{S/T}$ is due to electron-electron interactions and is influenced by the dielectric constant κ rather than the hopping element t. For $t = -4.2$ eV, $\Delta_{S/T}$ varies from 0.15 eV for $\kappa \sim 6$ to 0.35 eV at $\kappa \sim 2$.

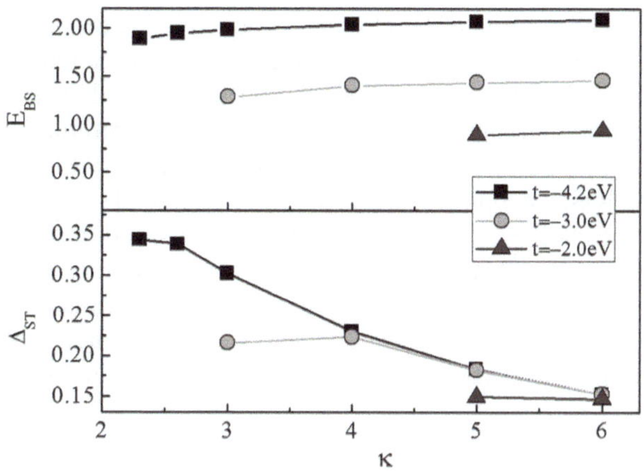

Fig. 7.15 Position of the bright degenerate singlet and the bright-singlet–dark-triplet separation as a function of κ

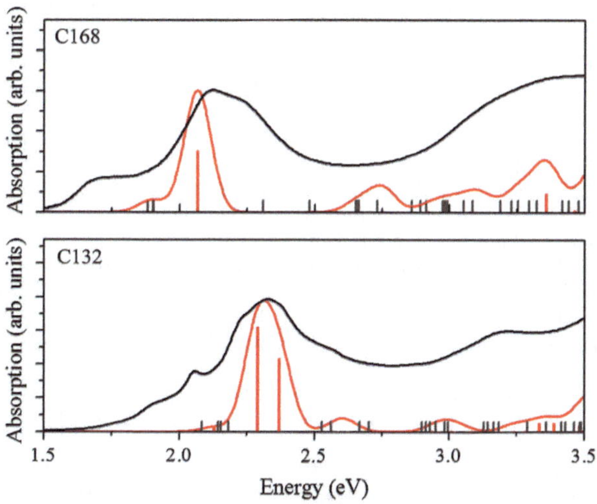

Fig. 7.16 Absorption spectrum of C168 *upper spectrum* and C132 *lower spectrum* compared with the experiment at $T = -4.2\,\text{eV}$, $T_2 = -0.1\,\text{eV}$ and $\kappa = 5$. 10 % of the highest absorption peak has been assigned to absorption of dark singlets. The *red straight line* is the calculated absorption. *Red drop lines* are singlet absorption peaks, *gray drop lines* represent the location of triplets and the *black straight line* is the experimental absorption data. Reprinted from [20]

7.5.5 Comparison With Experiment

We now compare the calculated absorption spectra with experiment. Figure 7.16a shows the measured [17] and calculated absorption spectra for $\kappa = 5.0$ and $t = -4.2\,\text{eV}$. We have used Gaussian broadening in continuous plots and added 10 % of the oscillator strength of the brightest peak to the dark singlets since they may contribute to absorption if the symmetry is broken due to, e.g., charge and spin fluctuations in the surrounding fluid. We see that the measured absorption spectra show an absorption threshold around $E = 1.8\,\mathrm{eV}$, a peak at $E = 2.25\,\mathrm{eV}$ and a reduced absorption strength up until $E = 3\,\mathrm{eV}$. Our preliminary interpretation assigns the peak in the measured absorption spectrum at $E = 2.25\,\mathrm{eV}$ to the bright singlet excitons while we predict the absorption threshold as due to dark singlets which dictates the choice of t and κ. The calculated absorption spectrum can reproduce the position of the absorption peak due to bright excitons followed by a gap. However, the singlet/triplet splitting is significantly underestimated when compared with experiment. More work is needed toward the understanding of the electronic structure and optical properties of graphene quantum dots, including the effects of impurities [26].

References

1. J. Peng, W. Gao, B.K. Gupta, Z. Liu, R. Romero-Aburto, L. Ge, L. Song, L.B. Alemany, X. Zhan, G. Gao, S.A. Vithayathil, B.A. Kaipparettu, A.A. Marti, T. Hayashi, J. Zhu, P.M. Ajayan, Nano Lett. **12**, 844–849 (2012)
2. S. Kim, S.W. Hwang, M.-K. Kim, D.Y. Shin, D.H. Shin, C.O. Kim, S.B. Yang, J.H. Park, E. Hwang, S.-H. Choi, G. Ko, S. Sim, C. Sone, H.J. Choi, S. Bae, B.H. Hong, K. Hee, ACS Nano **6**(9), 8203–8208 (2012)
3. S. Kim, D.H. Shin, C.O. Kim, S.S. Kang, S.S. Joo, S.-H. Choi, S.W. Hwang, C. Sone, Appl. Phys. Lett. **102**, 053108 (2013)
4. J. Akola, H.P. Heiskanen, M. Manninen, Phys. Rev. B **77**, 193410 (2008)
5. Z.Z. Zhang, K. Chang, F.M. Peeters, Phys. Rev. B **77**, 235411 (2008)
6. K.A. Ritter, J.W. Lyding, Nat Mater. **8**, 235 (2009)
7. A.D. Güçlü, P. Potasz, P. Hawrylak, Phys. Rev. B **82**, 155445 (2010)
8. M. Byszewski, B. Chwalisz, D.K. Maude, M.L. Sadowski, M. Potemski, T. Saku, Y. Hirayama, S. Studenikin, D.G. Austing, A.S. Sachrajda, P. Hawrylak, Nat. Phys. **2**, 239 (2006)
9. D.M. Hoffman, P.C. Eklund, R.E. Heinz, P. Hawrylak, K.R. Subbaswamy, Phys. Rev. B **31**, 3973 (1985)
10. A. Wojs, P. Hawrylak, Phys. Rev. B **55**, 13066 (1997)
11. J. Blinowski, N.H. Hau, C. Rigaux, J.P. Vieren, R. Le Toullec, G. Furdin, A. Herold, J. Melin, J. Phys. **41**, 47 (1980)
12. F. Wang, Y. Zhang, C. Tian, C. Girit, A. Zettl, M. Crommie, Y.R. Shen, Science **320**, 206 (2008)
13. Z.Q. Li, E.A. Henriksen, Z. Jiang, Z. Hao, M.C. Martin, P. Kim, H.L. Stormer, D.N. Basov, Nat. Phys. **4**, 532 (2008)
14. A.D. Güçlü, P. Hawrylak, Phys. Rev. B **87**, 035425 (2013)
15. X. Yan, X. Cui, L. Li, J. Am. Chem. Soc. **132**, 5944 (2010)
16. X. Yan, X. Cui, B. Li, L. Li, Nano Lett. **10**, 1869 (2010)
17. M.L. Mueller, X. Yan, J.A. McGuire, L. Li, Nano Lett. **10**, 2679 (2010)
18. X. Yan, B. Li, X. Cui, Q. Wei, K. Tajima, L. Li, J. Phys. Chem. Lett. **2**, 1119 (2011)
19. X. Yan, B. Li, L. Li, Accounts of chemical research (2012). doi:10.1021/ar300137p
20. I. Ozfidan, M. Korkusinski, A.D. Güçlü, J. McGuire and P. Hawrylak, Phys. Rev. B (2014, in press)
21. A. Trojnar, M. Korkusinski, E. Kadantsev, P. Hawrylak, Phys. Rev. B **84**, 245314 (2011)
22. S. Sorella, E. Tosatti, Europhys. Lett. **19**, 699 (1992)
23. T.O. Wehling, E. Sasioglu, C. Friedrich, A.I. Lichtenstein, M.I. Katsnelson, S. Blügel, Phys. Rev. Lett. **106**, 236805 (2011)
24. A.H. MacDonald, J. Jung, F. Zhang, Phys. Scr. **T146**, 014012 (2012)
25. P. Potasz, A.D. Güçlü, A. Wójs, P. Hawrylak, Phys. Rev. B **85**, p. 075431 (2012)
26. H. Riesen, C. Wiebeler, S. Schumacher, J. Phys. Chem. A. **118**, 5189 (2014)

Index

A.D. Güçlü et al., *Graphene Quantum Dots*,
NanoScience and Technology, DOI 10.1007/978-3-662-44611-9

MIX
Papier aus verantwortungsvollen Quellen
Paper from responsible sources
FSC® C105338

If you have any concerns about our products,
you can contact us on
ProductSafety@springernature.com

In case Publisher is established outside the EU,
the EU authorized representative is:
Springer Nature Customer Service Center GmbH
Europaplatz 3, 69115 Heidelberg, Germany

Printed by Libri Plureos GmbH
in Hamburg, Germany